AIR COMMAND AND STAFF COLLEGE

AIR UNIVERSITY

MOVEMENT AND MANEUVER IN DEEP SPACE:
A Framework to Leverage Advanced Propulsion

by

BRIAN E. HANS, Major, USAF
CHRISTOPHER D. JEFFERSON, Major, USAF
JOSHUA M. WEHRLE, Major, USAF

A Research Report Submitted to the Faculty

In Partial Fulfillment of the Graduation Requirements for the Degree of

MASTER OF MILITARY OPERATIONAL ART AND SCIENCE

Advisor: Lt Col Peter A. Garretson

Maxwell Air Force Base, Alabama

April 2017

Disclaimer

The views expressed in this academic research paper are those of the authors and do not reflect the official policy or position of the US government or the Department of Defense. In accordance with Air Force Instruction 51-303, it is not copyrighted, but is the property of the United States government.

Abstract

This analytical study looks at the importance of Deep Space Operations and recommends an approach for senior policy leaders. Section 1 presents a capability requirements definition with candidate solutions and technology strategies. Section 2 recommends an acquisition and organizational approach. Section 3 provides an extended strategic rationale for deep space operations as a national priority.

Contents

Illustrations

Tables

6

Introduction

This chapter of Air University's Space Horizons Research Group presents capability requirements, potential solutions, and strategic rationale for achieving movement and maneuver advantage in deep space. In this context, *deep space* is anything beyond geosynchronous Earth orbit (GEO). Driving the research are two primary assumptions underpinning the need for investment in deep space propulsion. The first assumption is that growing international activity, commerce, and industry in space extends the global commons, thus creating a military-economic imperative for the United States Department of Defense (DoD) to expand its protection of U.S. interests by defending space lines of communication. Although there are wide-ranging reasons to expand the space-faring capabilities of the human species, from the capitalistic to the existential, the fact of its occurrence offers the U.S. immense strategic opportunity. Section 1, operating on this assumption, recommends capability-based requirements for deep space operations given a projected future operating environment.

The second driving assumption underpinning this study is that improved movement and maneuver capabilities in deep space offer a wide array of benefits for the current National Security Enterprise, and for this reason alone demands attention in the form of disciplined investment. Furthermore, because the core functional capability required for deep space operations is in-space propulsion, the requirement necessitates a materiel solution. Although there are significant implications for the other DOTMLPF elements (e.g., requisite changes to Doctrine, Organization, Training, Leadership and Education, Personnel, and Facilities), they are not addressed by this study. Section 1.1, operating on the above assumption, highlights advanced and potential breakthrough propulsion technologies as candidate solutions to address the capability gap. Mach Effect Thrusters and EmDrive emerge as the most enticing potential breakthroughs because they offer virtually free thrust in exchange for electricity and are relatively inexpensive to investigate, yet much remains unproven. Section 1.2 continues by presenting two complementary approaches to the assessment of candidate technologies and pioneering research, along with cost implications. Section 1.3 closes with the strategic opportunity offered by placing advanced propulsion within a chain-link system of systems, resilient to hacking, replication, or leapfrogging.

Section 2 of this study recommends a two-part solution for acquiring deep space propulsion capabilities, fully acknowledging the parallel requirement to create affordable access to space. Section 2.1 involves a brief analysis of current efforts by the DoD and USAF to streamline acquisition timelines, followed by a proposed acquisition model to develop and deploy deep space propulsion technologies while collaborating with agencies and organizations external to the USAF. Section 2.2 discusses a theoretical organization formed and chartered to develop, test, and acquire deep space propulsion technology and includes what the organization would potentially look like. Finally, Section 3 provides an extended strategic rationale for deep space propulsion to close the study with further elucidation of the underlying imperative. The entirety of this work provides decision-makers a framework to identify and leverage advanced propulsion technologies to enhance Joint Force capabilities in deep space, in particular to achieve movement and maneuver advantage.

Section 1 – Capability Requirements Definition

"The Air Force needs to focus on true "strategic" objectives in space. These are objectives for the coming Century... True space operations will spread across the solar system in the decades ahead and the nation that controls them will dominate the planet. Focusing on LEO is akin to having a Navy that never leaves sight of the shore. The US Military needs to focus on "blue-water" space operations – GEO and above. US military space operations need to be in deep space, initially all of cislunar space, with an eye upon the entire inner solar system. To operate in deep space one needs to use the resources there, starting with fuel from asteroids. Once this is recognized, the military-economic imperative of identifying and protecting these assets becomes clear. The focus... should be to be sure on low-cost access to real outer space – with "space" beginning at GEO. New means of moving around in space are more important than just getting off the ground."

- Brigadier General (Retired) S. Pete Worden, USAF[1]

The Space Warfighting Construct (SWC) has reoriented the U.S. National Security Space Community toward improvements in resiliency, operations, and force presentation to the Joint Force Commander across all space mission areas. Although the SWC successfully reprioritized current and future investments in treating space as a warfighting domain, the nation is at risk of falling behind in the development of national power in space commensurate with the ambitions of private industry and peer competitors. In the commercial space industry, capital and capability are reaching an expansion threshold for the creation of a cislunar marketplace in which tourism, lunar real estate, and access to resources from other near-Earth objects are the primary commodities. Meanwhile, NASA's Asteroid Redirect Mission (ARM) created requirements advancing deep space propulsion, proximity operations, and noncooperative capture and deflection—all requisite capabilities of a Joint Force operating in deep space. Finally, the opportunistic policies, intent, and actions of space-faring peer competitors such as China, Russia, and India, along with civil and commercial endeavors in space suggest a future operating environment in which:

- *More than one nation mines and moves asteroids*
- *More than one nation mines material from the lunar surface*

- *Nations either can, or have appetites to build solar power satellites in GEO and large habitats or depots at Lagrange Points in cislunar space*
- *There is a desire by both industry and nations to create a hydrogen-based economy*
- *There is a desire by both industry and nations to move manufacturing off-Earth*
- *There is a desire by both industry and nations to become a multi-planetary species*
- *The space economy is growing at a rate that it might eclipse the total terrestrial GDP*
- *Nations must police their own commerce and may be deputized to police others, or asked to behave in hostile manners toward other nation's activities (visit, board, search, seize)*

As the Air Force Future Operating Concept states, "Continued expansion into space and cyberspace will increase the magnitude of the Air Force's operating area. The Air Force of 2035 will continue to perform five core missions, but advanced technologies and approaches will extend their scope."[2] Simply putting "space" in front of the existing USAF core missions in Figure 1 merely reinforces the concept that deep space operations are an extension of the use of the military instrument of power on Earth. However, an expanded operating area in space carries its own challenges and opportunities, offering different ways and means of wielding the military instrument of power. Thus, the roles and missions of the United States Air Force in space are also subject to change. Furthermore, because space power theory is based on more than military capability alone, parallels are drawn to the U.S. Navy's political-economic role in fostering trade relationships while protecting sea lines of communication and overseas colonies throughout U.S. history. In his book, *Developing National Power in Space*, Dr. Brent Ziarnick adapts Mahan's sea power theory in making the case that space power is based on *access* and *ability* in that domain.[3] Therefore, one arrives at the conclusion that capabilities that enhance access and ability within space carry great potential to enhance national power. In military terms, the Joint Force requires the capability to project power in, through, and from the space domain. More pertinent to this study, *in-space propulsion* is the common functional requirement for the

USAF's current and future missions in space.

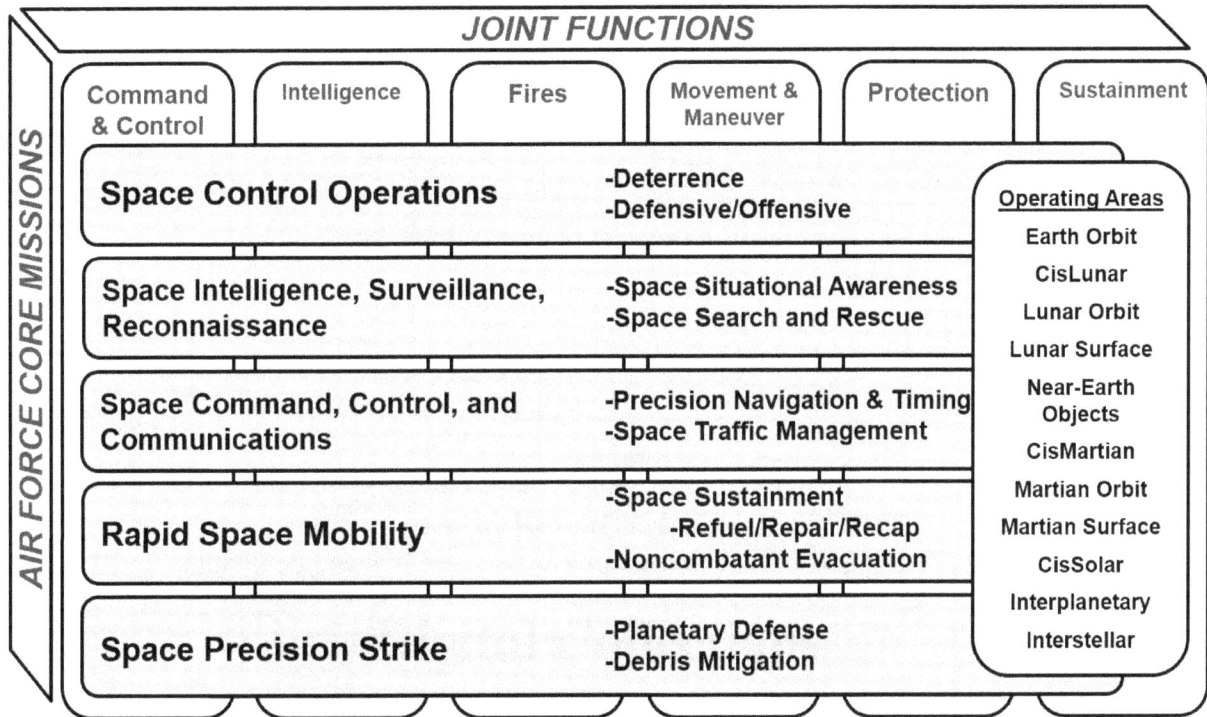

Figure 1. Notional USAF Core Missions in Space

Recommendation 1: based on a projected operating environment and notional set of missions,

the following requirements should be adopted by the Joint Requirements Oversight Council:

- **The Joint Force requires rapid mobility and maneuver including the inner solar system** to establish, use and protect space resources for energy production and transfer, including materials for manufacturing and maintenance, and to protect licit commerce.

- **The Joint Force requires propulsion systems capable of in-situ resource utilization (ISRU)** and transit within the inner solar system at least as fast as commercial vessels. Military operations are limited only by the capability of the propulsion system to respond in a timely manner, and the change in velocity it allows. Therefore the Joint Force must continually invest in propulsion systems that are higher specific impulse (I_{sp}), higher thrust, higher specific power, and better able to use resources in space.

- **The Joint Force requires a system of systems in space capable of reusable space-tugs, on-orbit refuelers, fuel depots, and spacecraft capable of receiving in-space fueling and servicing.** Development of very high I_{sp} and very high I_{sp}/high thrust engines and/or wireless power transmission are required which offer significant payload, speed, or endurance capabilities over chemical propulsion. This capability seeks a continuing maneuver advantage over an adversary attempting to control or deny the domain, and anticipates operations requiring significant maneuver in the cislunar system on the time-scale of days, and operations within the economic sphere of the inner solar system, including space operations against objects with significant mass.

Although a view of the USAF in 2035 was referenced above, it is important to clarify that this study looks toward the 22nd Century USAF, with intended benefits to near-term capabilities. In order to do that, one recommendation is to expand the Rocket Propulsion for the 21st Century (RP-21) program in order to drive breakthrough propulsion research and development. RP-21 is a coordinated effort between the DoD, NASA, and industry to develop revolutionary and innovative rocket propulsion technologies by the year 2027 according to quantified improvement goals. As developmental planning and communication tools, the AFRL technology roadmaps referenced in this study map technology efforts to both RP-21 goals and USAF technology needs. Furthermore, the AFRL roadmaps link technologies and development efforts to SWC attributes such as space superiority, resilience, and sustainment. For example, high thrust may translate to superior maneuverability, while long duration mission life and cycling translates to resilience. In addition to expanding the RP-21 program to drive propulsion breakthroughs beyond the near-term horizon, one could add "range, cargo capacity, serviceability, or interoperability" to the SWC attributes to further qualify superiority and resilience.

Recommendation 2: the USG, DoD, or USAF can expand the RP-21 program beyond 2027 in order to drive breakthrough propulsion research and development. Add deep space performance goals to the Space Warfighting Construct to further qualify superiority and resilience, such as "range, cargo capacity, serviceability, or interoperability."

While the attributes emphasized by the Space Warfighting Construct are important as the USAF moves into future missions in deep space, the USAF can also learn from NASA's approach to guiding research and technology development by using Design Reference Missions (DRMs). DRMs are a planning tool used for technology trade studies, to analyze the effect of different approaches to missions within a higher-level Design Reference Architecture (DRA). NASA's current mission planning architecture is DRA 5.0, Human Exploration of Mars.[4] Within DRA 5.0 lies the Asteroid Redirect Mission, which established requirements for advancing deep space propulsion, proximity operations, and non-cooperative capture and deflection. These requirements represent *proximate objectives* toward missions to Mars, in-situ resource utilization (ISRU), and planetary defense capabilities. A "proximate objective is guided by forecasts of the future, but the more uncertain the future, the more its essential logic is taking a strong position and creating options."[5] In this way, DRMs may drive the accomplishment of proximate objectives, even when the future is uncertain, or evolve to create their own standalone capability advantages.

DRMs may be developed through space superiority-type mission planning based on current and projected space activity, however, a note of caution is the importance of avoiding overly prescriptive technology solutions. The lure of designing a mission around a particular technology is strong, but the initial focus needs to center on deriving *capability-based* requirements in order to prioritize technology development and pioneering research in applied science. This study offers two recommended approaches, one based on optimizing existing capabilities, and another directed toward breakthrough capabilities. An additional benefit to developing DRMs is their catalyzing effect for other non-materiel requirements to prepare for deep space operations. Conversely, one result of the mission analysis process may be a decision that a particular mission is best addressed in a different way or by another entity. Will it be the USAF's role to conduct search and rescue operations in space or does that risk continue to be accepted and designed into human spaceflight? In this way, the *process* of researching, analyzing, and refining DRMs is the most important first step to planning for future space operations.

Recommendation 3: based on anticipated operations in deep space, the USAF can succinctly capture *capability-based requirements* by developing Design Reference Missions (DRMs)—within reference architectures—to drive meaningful scientific and technology work. DRMs should be developed by multidisciplinary teams comprised of subject matter experts across government, industry, and academia—with combined experience in concept design, engineering, and operations. DRMs and corresponding capability requirements should be developed in enough detail to guide scientific research and technology development, as well as solutions based on the non-materiel requirements of those missions (e.g. DOTMLPF implications).

Section 1.1 – Technology Survey

"Markets that do not exist cannot be analyzed: Suppliers and customers must discover them together. Not only are the market applications for disruptive technologies unknown at the time of their development, they are unknowable. The strategies and plans that managers formulate for confronting disruptive technological change, therefore, should be plans for learning and discovery rather than plans for execution."
- *Clayton M. Christensen, The Innovator's Dilemma[6]*

Regardless of the shape, order, or composition of the future operating environment, current trends project a Joint Force capability gap in deep space, not only in-space movement and maneuver, but a chain link of capabilities including deep space ISR, C2, PNT, search and rescue, sustainment, debris mitigation, and, if necessary, planetary defense (because NASA is currently charged with the mission). Following Christensen's advice from *The Innovators Dilemma*, the U.S. Government can prepare for an uncertain future in deep space by seeking that which needs to be known (what he calls *discovery-driven planning*).[7] The cornerstone of that work is underway via the RP-21 program and other government-funded entities such as ARPA-Energy with respect to harnessing nuclear fusion. However, those disparate efforts lack the objective focus consistent with government-industry initiatives on the scale of trans-continental railroads or overseas trade. From the standpoint of in-space propulsion, narrowing the projected capability gap starts at both ends: working forward from existing technologies while working backward from experimental concepts on the near-frontiers of propulsion science. Therefore, a two-pronged approach is recommended: 1) assessing current technologies and their potential improvements based on capability-based measures of performance via Design Reference Missions, and 2) assessing breakthrough propulsion candidates for their potential gains (in mass, speed, or energy) versus cost (in resources and time) according to basic principles of experimental rigor. Because this study is limited in time and scope, the two-pronged approach

15

mitigates risk of losing first-mover advantage by *not* pursuing a particular concept, and serves as a model for future capability-based assessment.[*]

Technologies

There are many candidate technologies for in-space propulsion systems. Presented here, in brief, are the most relevant to deep space operations based on their current or potential capabilities and the recommendations of experts interviewed for this study. By no means is the list all-inclusive, but it serves as a foundation for further investigation. Only a sampling of performance measures are included in order to provide context for comparison. In general, performance goals for advanced propulsion include improvements in thrust levels, specific impulse (I_{sp}), power, specific mass (or specific power), volume, system mass, system complexity, operational complexity, commonality with other spacecraft systems, manufacturability, durability, safety, reliability, and cost.[8]

Chemical Propulsion – Advanced Cryogenic Evolved Stage (ACES)

Chemical propulsion is a mature solution with plenty of flight heritage, and worth mentioning because it is sustainable under an assumed architecture of extra-planetary resource mining and propellant manufacture, otherwise known as in-situ resource utilization (ISRU, e.g. asteroid mining, energy production, and fuel depots). As an example, the United Launch Alliance (ULA) is developing the Advanced Cryogenic Evolved Stage (ACES) to be the

---

[*] Millis outlines three basic risks to revolutionary research and their corresponding mitigations on page 698. For comprehensive alternative approaches to technology assessment, see also Funaro, Gregory V., and Reginald A. Alexander. "Technology Alignment and Portfolio Prioritization (TAPP): Advanced Methods in Strategic Analysis, Technology Forecasting and Long Term Planning for Human Exploration and Operations, Advanced Exploration Systems and Advanced Concepts." (2015).

workhorse of the cislunar economy. Intended to replace the Centaur second stage in 2023 with at least four times its thrust (450-650kN), ACES will use existing hydrogen (H2) and oxygen (O2) propellants in a new Integrated Vehicle Fluids (IVF) system. The IVF system reduces weight and complexity by combining the electrical power, chemical pressurization, and reaction control systems, burning waste hydrogen and oxygen from the main engines. Other attributes of ACES include full in-space reusability, unlimited engine starts, and a 68,000 kg propellant load, which enables mission durations from hours to weeks depending on the profile, or indefinitely given in-space propellant depots.[9] ACES represents one option among many chemical systems available for the first reusable, extended, bulk mobility operations in space.

Viable chemical propulsion systems for deep space will use either cryogenic or storable propellants. By using H2 and O2, ACES avoids using the high-performing but unstable and highly-toxic hydrazine-based propellants. Of note, NASA selected the Air Force-developed AF-M315E for flight demonstration as the highlight of their green propellant infusion mission (GPIM). AF-M315E is intended to replace hydrazine across mission classes (in the 1-to-22N thrust range) due to its high performance and very benign safety properties.[10] Because chemical systems still offer a mature solution for high performance, on-demand thrust, AFRL continues work on next-generation non-toxic, high-density propellants to enhance time-constrained impulse maneuvers as a key attribute for superiority and resiliency under the SWC.[11]

Nuclear-Thermal Propulsion (NTP) and Nuclear-Electric Propulsion (NEP)

Many of the experts interviewed for this study recommend further pursuit of nuclear sources of rocket thrust, specifically the spectacular energies created by nuclear fusion. While Radioisotope Thermal Generators (RTGs) have flown on spacecraft since 1961 to generate electrical power as a type of Nuclear Electric Propulsion (NEP), they are radically different from Nuclear Thermal Propulsion (NTP). Elements within NASA recommend *fission* technologies (used on Earth for over 70 years) for deep space missions in order to reduce transit times for larger vehicles. Fission is the process of splitting an atom, releasing immense heat energy into a propellant, and then accelerating it through a nozzle (thus categorization as NTP). A propellant, typically hydrogen, produces thrust directly related to the thermal power of the reactor, while I_{sp} is directly related to exhaust temperature, ranging between 830-1000 seconds. This is an improvement over chemical rockets due to the lower average molecular weight of the propellant. NASA Marshall Space Flight Center maintains an NTP project capable of scaling to full demonstration in three-to-four years leveraging existing infrastructure at a cost of $1-1.5 billion.[12] Not only is it cheap compared to aircraft engine development costs, the project manager assesses that an NTP demo would mean 26,000 job-years across 28 states and could be accomplished within the current administration.[13] Alternatively, a lower-risk, lower-cost "zero-power-critical" engine demo (which operates at only one watt with no hydrogen propellant flow) could be accomplished for $150-200 million under three years. Such a demo would produce spin-off benefits to STEM education and infrastructure while the requisite Nuclear Regulatory Commission issues are resolved in parallel.[14]

Nuclear *fusion*, in contrast, heats a gas to separate its ions and electrons until they overcome their mutual repulsion and fuse together, releasing about one million times more energy than a chemical reaction and three-to-four times more than a fission reaction for equivalent mass.[15] For an excellent depiction of the potential range of nuclear thermal propulsion capabilities, see the Project Rho website.[16] Of note, U.S. Navy funding (about $18 million over a combined 10 years between 1992-2006) of Dr. Robert Bussard's Polywell inertial-electrodynamic fusion (IEF) device achieved a record deuterium-deuterium fusion output in final experiments.[17] Before Dr. Bussard passed away in 2006, a peer-reviewed report to International Astronautical Congress in 2006 reported that "Design studies of IEF-based space propulsion (*AIAA Prop. Conf, 1993,97; IAC, Graz, 1994, Toulouse, 2001*) show that this can yield engine systems whose thrust/mass ratio is 1000x higher for any given specific impulse (I_{sp}), over a range of $1000 < I_{sp} < 1E6$ sec, than any other advanced propulsion means, with consequent 100x reduction in costs of spaceflight."[18] Full scale net-power demonstration would require \$180-200 million over five years, depending on the fuel combination selected. Bussard's pioneering work in IEF represents one of the best candidates for space propulsion applications, but there are others.

In the prospective clamor for position in an energy industry revolution, there are a number of fusion start-up development efforts worth investigating. These are in addition to the internal R&D of large defense contractors like Lockheed Martin (LM). LM Skunk Works is developing a compact fusion reactor intended to reduce in-space transit times, among a host of complementary energy production objectives.[19] As one interviewee noted, "the fact that this is now done mostly because of private investment shows how far the Government has gone in being risk-adverse." The primary advantage of a nuclear system is much higher energy and power density than chemical propulsion. Furthermore, nuclear energy can be used in thermal or electric propulsion systems, such as the VASIMR system described next. As the interviewee noted, "there is nothing here that fundamentally violates known laws of physics, and it is really a question of engineering and scaling." Finally, nuclear propulsion offers immense capability on its own, made more valuable without the assumption of in-situ resource utilization or beamed-power architectures that may be unavailable in the near future.

Variable Specific Impulse Magnetoplasma Rocket (VASIMR)

VASIMR is a high-power radio frequency driven plasma thruster capable of I_{sp}/thrust modulation at constant input power—meaning it can "shift gears" from about 3,000 seconds I_{sp} to 30,000 seconds by trading off thrust. It would use a low gear to climb out of planetary orbit, and high gear for interplanetary cruise. Other advantages of VASIMR are increased lifetime (due to removal of the electrodes normally present in electric propulsion technologies), heating efficiency, and a mass-saving power conditioner.[20] The VX-200 engine is a VASIMR prototype developed by former astronaut Franklin Chang-Diaz and his Ad Astra Rocket Company. The VX-200 has been "tested under space-like conditions as a technology demonstration and risk mitigation platform, in addition to serving as a means to explore fundamental plasma physics for academic purposes."[21] NASA's roadmaps maintain VASIMR at Technology Readiness Level 3 (TRL 3 = analytical and experimental proof of concept[22]) and a "near-term objective is maturation of a 30 to 200 kW-capable dual thruster system to flight demonstration for solar-powered cislunar space tug operations, and exploration to Mars and Jupiter's icy moons." Finally, VASIMR has the benefit of chemical or nuclear-level thrust profiles without similar risks of pollution.

Directed Energy-Driven Technology

Just as in-situ resource utilization (ISRU) enables sustainable chemical propulsion in space, an alternate or complementary architecture is space-based solar power (SBSP). The technology that enables beaming solar energy to Earth would also enable wireless energy transmission in space. Beamed energy propulsion (BEP) transmits laser or microwave energy from a ground or space-based energy source to an orbital vehicle, which uses it to heat a propellant or reflect beamed energy to generate momentum.[23] The beams can be continuous or pulse mode and provide more flux intensity than sunlight. The advantage is a propulsion system capable of delivering low thrust with high I_{sp}. In addition, wireless power transmission can reduce spacecraft weight via on-board, modular, or close proximity systems.[24] Although beam control, pointing, and tracking systems present a major challenge in this high-power domain, the technology can also be used for orbital debris removal with laser ablation.[25] Note that high power levels are required for fast, inexpensive, and long duration mobility and maneuver in space.

While the field of directed energy is relatively mature for terrestrial applications, space applications are not. However, researchers from the terrestrial side concluded at a workshop in early 2016 that no fundamental technological obstacles prevent moving BEP from laboratory research to space-based applications.[26] Furthermore, Dr. Philip Lubin and fellow researchers proposed an orbital platform called DE-STAR for Directed Energy Solar Targeting of Asteroids and exploRation. DE-STAR is a modular phased array of lasers, powered completely by solar technology. It is designed as a multi-tasking system capable of many different uses when not in use for its main objective of defending Earth by using focused directed energy to raise the surface spot temperature of an asteroid to ~3,000 K, allowing direct evaporation of all known substances.[27] Additionally, DE-STAR can be used as a LIDAR system to detect asteroids, as a photon drive to propel spacecraft up to relativistic speeds, as a mining system to analyze the compositions of various asteroids and celestial bodies, and as a communications array.[28] While the concept is futuristic, many of the core technologies currently exist and small, inexpensive systems can be built to test the basic concepts on a modular prototyping path.[29]

Solar Thermal Propulsion

Leading an economic analysis of space transportation architectures supplied from near-Earth object (NEO) resources, Dr. Joel Sercel of TransAstra Corporation found that using directed solar energy to heat propellant would reduce infrastructure costs and enable a cislunar mining economy.[30] Made available in cislunar space, water can be used directly as propellant in Solar Thermal Rockets (STRs) to provide inexpensive transportation. STRs are estimated to provide about 365 sec I_{sp}. Dr. Sercel's Omnivore engine is a promising solution for in-space propulsion as a "flex fuel" solar thermal system capable of using nearly any fluid as propellant, nominally "dirty" water from asteroids.[31] As a complementary capability, TransAstra developed an optical mining testbed in collaboration with the Colorado School of Mines, which simulates highly concentrated sunlight to drill holes, excavate, disrupt, and shape an asteroid's surface.[32] Optical mining of NEOs is synergistic with solar thermal propulsion because both require concentrated sunlight and center on water-based propellants.[33] TransAstra conducted a study using a variety of reference missions and spacecraft configurations within three different architectures—one without the assumption of ISRU and two ISRU-variants (water-only and liquid oxygen-methane). Merely for illustration, even if conservative cost-of-launch estimates remain the same over 20 years ($75,000 per kilogram), non-ISRU total costs are over $310 billion, while ISRU costs are less than $91 billion.[34] Ultimately, TransAstra's goal is to reverse the spiral of cost growth in space transportation, which would enable a thriving deep space economy for the nation.

Electric Propulsion – Hall Effect, Field Reverse Configuration, and Electrospray Thrusters

Electric propulsion generally ionizes a propellant and accelerates the ions (or resultant plasma) in the opposite direction of desired motion according to conventional rocket science. The primary advantage of electric propulsion is much higher I_{sp} (due to electrical efficiency), although at the expense of high thrust. Over long durations, however, large changes in velocity (delta-V) are possible, as are multiple restarts. Another advantage to electric propulsion is that technical challenges can be solved by engineering (e.g. "scaling-up"), rather than dealing with issues of fundamental physics, as in the case with potential breakthroughs described below.[35] Electric propulsion is a broad field, ripe with opportunity to engineer improvements in not only thrust, efficiency, duration, and sustainability, but other unique attributes such as multi-mode operations, "flex-fuel" engines, signature reduction, and thrust vectoring.

AFRL is pursuing the attributes listed above in different forms, particularly via multi-mode propulsion where both chemical thrusters and electric propulsion devices operate on a common propellant for operational flexibility. Called field reverse configuration (FRC) thrusters, they use the interaction between induced plasma currents and applied magnetic fields to accelerate plasma.[36] With low mass and efficiency comparable to Hall thrusters, the unique advantage of FRCs is their ability to use almost anything as propellant. Finally, electrospray thrusters, also in development at AFRL, are very low mass devices that accelerate a range of ions at high efficiency. Their unique attributes include small applications (e.g. formation flying), thrust vectoring via phased array-like grids, and ability to scale up indefinitely. Electric propulsion's inherent flexibility and scale-up potential offer a wide range of candidates for advanced in-space propulsion, foremost of which are ion propulsion systems such as Hall Effect Thrusters, FRCs, and electrosprays.

Of note, the potential to scale the Hall Effect Thruster (HET) for higher thrust and efficiency led AFRL and NASA to devote considerable resources to maturing the technology.[37] The Hall effect is a force that results when ions in a plasma are accelerated via cross-field discharge—the interaction between a radial magnetic field and the electric field induced by its application to a conducting plasma.[38] NASA matured a 12.5 kW Hall Effect Rocket with Magnetic Shielding (HERMeS) to TRL 6 (prototype demonstration in a relevant environment) and transferred it to the U.S. commercial market. Companies then competed for the spaceflight application contract, won by Aerojet Rocketdyne as part of the Asteroid Redirect Robotic Mission (ARRM).[39] ARRM is the first sub-mission within the overarching asteroid redirect DRM. This method of technology maturation and transfer serves as a promising model for commoditizing advanced in-space propulsion technology.

Boeing Phantom Works competed for the ARRM flight integration contract, and as leaders in high power solar-electric propulsion (HP-SEP), will be prime candidates in the marketplace going forward. Boeing transitioned Xenon-Ion Propulsion Systems (XIPS) into the 702 line of satellite buses, included an all-electric 702-SP.[40] As some engineers note, the improvement in HP-SEP technology required by ARRM provides an extensible path to a range of new and enhanced missions in civil and military space, including affordable removal of large orbital debris objects.[41] The commoditization of electrical propulsion (although applications remain proprietary), along with advanced chemical options such as ULA's ACES, provide a prospective foundation for current and future deep space operations.

Mach Effect Thrusters (MET)

Airmen are likely familiar with *Mach number* as a ratio of a body's speed to the local of sound, yet unfamiliar with Ernst Mach's theories regarding space, time, and matter. Einstein defined *Mach's principle*, the foundation for *Mach effects*, as the "relativity of inertia," such that a material object's inertial properties are related to the presence and action of surrounding objects (the entire universe).[42] However, there is very little consensus on the source of inertia itself—not to mention gravity—and debates abound. Physicist James Woodward, the inventor of the Mach Effect Thruster, devotes an entire book to the mathematical derivation and description of Mach effects, which are "predicted fluctuations in the masses of things that change their internal energies as they are accelerated by external forces."[43] Woodward created a device that produces micronewton-level thrust by changing the size and shape of a piezoelectric material at high frequency (and thereby its mass in relation to the universe around it) by applying sinusoidal voltage to it.[44] The Mach effect device relies on an understanding of inertia as the cumulative gravitational interaction of all the mass-energy in the universe, and peer reviewers now agree on the General Relativity Theory (GRT) underpinnings—but not the mathematical derivation—of the net impulse as a product of timing the mass fluctuations with internal constituent motions of the thruster.[45]

An independent assessment by the Aerospace Corporation on behalf of the USAF concluded in 2014 that Mach effect research at the time was not sufficiently mature to warrant government funding.[46] However, laboratory-controlled experiments using variations of the Mach effect device since that time have confirmed the presence of a "non-zero thrust signal," while consensus has arisen over the GRT underpinnings.[47] Although the observed force is very small, the implications of such propellantless breakthroughs upon space travel are revolutionary. At current spacecraft power levels, Mach effect thrusters could enable missions unconstrained by conventional propulsion system limitations such as mass, volume and consumption rate.[48] Researchers express a strong desire to scale-up experiments in order to move beyond quibbling over sources of experimental error or "noise."[49] The goal would be unambiguous evidence of a new physical effect to the scientific community and public writ large, but obvious institutional and funding barriers stand in the way. As a starting point, Woodward recommends increasing the sinusoidal voltage frequency from kilohertz to megahertz using automatic frequency control, although materials science and engineering work would be required to produce new piezoelectric materials and compensate for natural resonance, mechanical fatigue, and thermal effects.[50] Developing such thrusters "that operate large volumes of inertia-varying mass at high frequencies could, in principle, produce macroscopic thrust and lift."[51]

EmDrive

There are three things USAF leadership needs to know about EmDrive (short for electromagnetic drive). First, as a propellantless source of thrust, if proven it could revolutionize not only in-space transportation, but also an unexplored amount of Earth-based applications. Second, current experimental results are highly suspect due to potential sources of error and apparent violation of classical physics principles. Third, although unproven, the Chinese have embraced its potential with claims of laboratory evidence and even an alleged on-orbit experiment.[52] If the third point does not provoke a competitive instinct among airmen, the Naval Research Lab (NRL) is also attempting to reproduce and verify the latest results from NASA.

EmDrive thrust appears to result from high-frequency radio wave (RF) resonance within an asymmetrically shaped microwave cavity in vacuum, although it has no less than four theoretical explanations. As quantum theory goes, the empty space within the cavity is not actually empty—it is teeming with quantum field fluctuations. The RF resonance reduces quantum field fluctuation within the cavity such that an external force is detected—known as the *Casimir force*—which is analogous to a pressure imbalance created by a reduction in air density (think Bernoulli's principle).[53] Because the force appears to result from fluctuations

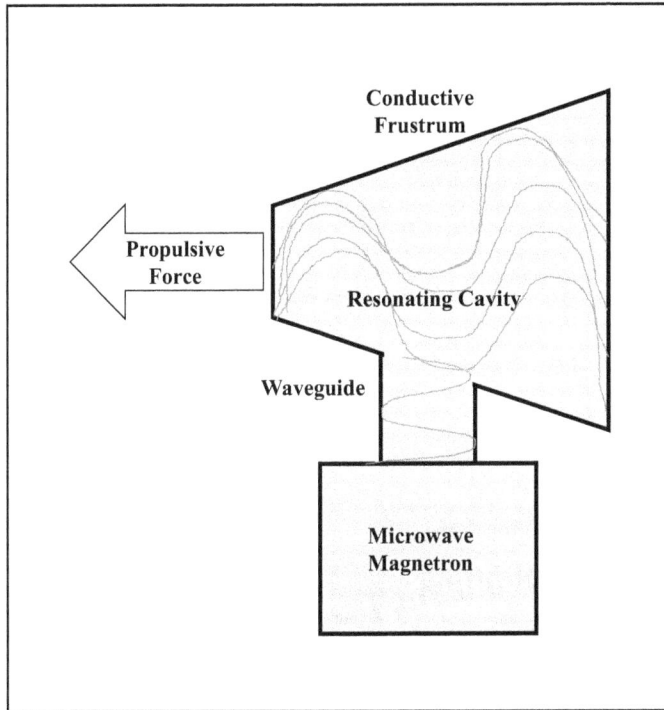

Figure 2. Typical EmDrive Experiment Configuration

in the quantum vacuum and not the electromagnetic (EM) effects, many see "Em" as a misnomer. At least this is one area where scientists agree: no thrust should be possible from EM effects alone. NRL researcher Mike McDonald places the EmDrive theories into two camps: "either the cavity pushes on something we didn't know was available to push on (quantum vacuum, dark energy, dark matter, the 'ether'), or else it somehow varies the inertial mass of some piece of the system in phase with the RF cycle, such that it pushes one way when a little 'heavier', one way when a little 'lighter.'"[54] Both camps contain "hopeful skeptics" who want to focus on getting the test right.

Recent fervor surrounding EmDrive followed reports in late 2016 from China and NASA Eagleworks' latest peer-reviewed article, "Measurement of Impulsive Thrust from a Closed Radio-Frequency Cavity in Vacuum." The Eagleworks experiment reported approximately 1.2mN/kW of thrust (which is about two orders of magnitude greater than the Mach effect experiments), but many researchers remain skeptical due to experimental sources of error. Even Dr. White and his Eagleworks team admit, "this test campaign was not focused on optimizing performance and was more an exercise in existence proof."[55] Scrutiny of the Eagleworks report is covered extremely well by a team of experts on the website Centauri-Dreams.org, including Marc Millis, who was the head of NASA's Breakthrough Propulsion Physics (BPP) project. They summarize the significant concerns about the Eagleworks experiment, which include thrust stand tilting, characterization, power cable forces, chamber wall interactions, and thermal effects. In sum, "the dynamic behavior of the thrust stand must be more thoroughly understood before reaching firm conclusions."[56]

As a recommended next step for the DoD, Travis Taylor of U.S. Army Space and Missile Defense Command offers an experimental verification methodology based on increased levels of EM input energy, up to and including coherent laser energy, which could theoretically produce thrust in the 30 N/kW range (see Table 1, below).[57] Combined with a rigorous but affordable experimental design on the order of $100-200,000 over 9-12 months (before analysis and review), Marc Millis and fellow researchers at the Tau Zero Foundation have the requisite background and expertise to move the research forward by creating reliable data.[58] The Aerospace Corporation, a Federally Funded Research and Development Center (FFRDC) also has the facilities, expertise, and credibility to accomplish such testing. DARPA expressed interest but remains in a holding pattern.

An important note about EmDrive and Mach effects: both represent avenues into understanding the nature of space-time, gravitation, inertial frames, quantum vacuum, and other fundamental physical phenomena, particularly the uncharted confluence of Quantum Field Theory and General Relativity Theory that gives rise to Quantum Gravity research.[59] Eric Davis and others suggest methods of "engineering the vacuum," envisioning the possibility that "an aerospace vehicle uses specially engineered energy fields to modify the local gravity field (by modifying the vacuum index of refraction) so that the craft can be lifted from the Earth's surface and propelled up to orbit. We can exploit this mechanism to propel an aerospace vehicle into and around space."[60] While it is premature to gauge the capabilities and limitations of EmDrive and Mach Effects, it is not too soon to dedicate resources toward further investigation. Although the research is relatively inexpensive, the science is a long way from weighing the energy costs—e.g. reducing the inertial mass of the spacecraft—against using that energy for conventional electric propulsion.[61] A difference worth noting between Mach effects and EmDrive is that the subject matter experts interviewed for this study agree that the demonstrated Mach effects are slightly "above the noise" of experimental error, whereas results from EmDrive experiments are still "below the noise."[62] To conclude, as Millis states in his review of the most recent results: "if either the EmDrive or Mach Effect Thruster is indeed genuine, then new physics is being discovered or old physics is being applied in a new, unfamiliar context. Either would be profound."[63]

Table 1: In-Space Propulsion Technology Summary

Propulsion	Thrust	Power	I$_{sp}$	Attributes
ACES	450-650 kN	unknown	unknown	68,000kg propellant; unlimited engine starts; hrs-to-wks mission duration; ISRU extends lifetime indefinitely
NTP-Fission	100+ kN	450 MW	1K sec	high thrust, I$_{sp}$ 2-3x solid core NTP, limited restarts, reduced initial mass in low-Earth orbit (IMLEO)
Adv. Fission	2000+ kN	GW	30-500K sec	extremely high I$_{sp}$, high-thrust, reduced IMLEO
NTP-Fusion	4-200+ kN	GW	10-100K sec	low mass, long-life, very high delta-V; fission-fusion hybrids reduce system mass by order of magnitude
VASIMR	900-5,900 mN	30-200 kW	2-5K sec	low mass, long-life, very-high delta-V, throttle-able
Hall Effect	13-252 mN	10-100 kW	1-3K sec	low mass, long-life, very-high delta-V, many restarts
XIPS	79-165 mN	2.2-4.5 kW	3.4-5K sec	low mass, long-life, high delta-V, many restarts
DE-STAR	333 N	50 GW	unknown	low-to-moderate thrust, low mass, high I$_{sp}$, multi-use
Mach Effect	0.01176 mN/kW	Propellant-less; theoretically scalable		
EmDrive	Propellant-less; theoretically scalable			
NASA-Eagleworks	1.2±0.1 mN/kW	Dr. White, et. al. Note that 1.2 mN/kW is over two orders of magnitude higher than other forms of "zero-propellant" propulsion, such as light sails, laser propulsion, and photon rockets having thrust-to-power levels in the 3.33–6.67 µN/kW (or 0.0033–0.0067 mN/kW) range (https://forum.nasaspaceflight.com/index.php?topic=36313.160)		
Shawyer	80-243 mN/kW	Inventor Roger Shawyer (http://emdrive.com/)		
China	290 mN/kW	Prof. Juan Yang et.al., 2012		
Taylor	3000 mN/kW	Laser-driven; proposed, 2017; equivalent thrust to a small fission reactor[64]		

Technology Survey Summary

In the course of this study, it became apparent that a side-by-side comparison between technologies would not be possible, nor would a meaningful analysis result from the absence of both detailed requirements and technical data within the resources and time available. NASA technologists echo this sentiment in their 2015 roadmaps, making no claim on a "one size fits all" solution. Rather, they conclude, "The development of higher-power electric propulsion, nuclear thermal propulsion, and cryogenic chemical propulsion will have the broadest overall impact on enabling or enhancing missions across each class [of *current* missions]."[65] Therefore, the following technology approaches will aid those involved in propulsion science and technology assessment, as well as future mission planners.

Section 1.2 – Propulsion Assessment Approaches: Existing and Breakthrough

Existing Technology: Using Measures of Performance and Design Reference Missions

Comparing propulsion technologies across categories (chemical, nuclear, electric, etc.) is precarious for two reasons. First, the performance envelopes vary, especially when projecting the outer bounds of incremental improvements still under development. Second, the relative importance of each measure changes according to the projected mission application and underlying assumptions. For example, the relative importance of high specific impulse (I_{sp}, or "bang for your buck") decreases if in-situ propellant resources are available. Although these are common engineering challenges, they increase uncertainty in the analysis. A comprehensive survey will involve multi-factor optimization and mitigate uncertainty with sensitivity analyses that account for differences in assumptions across DRMs.

This note of caution in comparing technologies echoes a fundamental issue in the propulsion science community. A recent call for topics by the Tau Zero Foundation's Tennessee Valley Interstellar Workshop (4-6 October 2017, Huntsville, AL) included the following:

> *Foundationally Consistent Baselines:* Different mission/vehicle concepts often use different projected performances for common functions such as: (a) heat rejection, (b) energy storage, (c) power management and distribution (PMAD), (d) magnetic nozzles, (e) communication with Earth, (f) equipment longevity, (g) structural mass {if built in space}, and (h) guidance, navigation and control (GNC). Fair comparisons of mission-vehicle concepts are difficult when different values are used for such baseline technologies. Presentations are invited that can credibly delineate reasonable performance estimates for such common functionalities so that future mission-vehicle studies can use common baselines for comparison (e.g. efficiencies, specific masses, readiness levels, etc).

> *Consistent Comparison Measures:* It is difficult to objectively compare different interstellar propulsion and power concepts that use different fundamental methods with method-specific performance measures (e.g. rocket specific impulse, laser pointing accuracy, etc). Abstracts are sought for suggested alternatives to compare both the abilities and resource requirements of diverse interstellar mission concepts – measures that are consistent across all modalities (perhaps in terms of energy, power, mass, mission time, etc.).[66]

Thus, while all of the above technologies have in common some degree of measurability, the methods and standards of measurement vary, just as their relative values depend on reference frames. For example, to which projected missions (or "modalities" in Tau Zero terms) are the above technologies best suited, and according to what attributes or measures of performance? While this circumstance seems rife with uncertainty, here is another opportunity to adapt and apply Christensen's lessons from *The Innovator's Dilemma* by approaching advanced propulsion technology development with a value-driven, capability-based method.

Step 1: for each propulsion technology or category, plot the S-curves of its performance measures. The measures of performance will lie on the vertical axis, while the horizontal axis represents cost (a function of time and engineering effort). Because these technologies are based on known physical principles and generally well-understood engineering, most if not all will asymptotically approach a projected performance limit, even if not yet demonstrated or costs are unknown. Note that in some cases, a design or operational trade-off exists, for example, a technology that "throttles" will trade thrust for efficiency, but the performance envelope is known or *knowable*.

Step 2: instead of trying to anticipate and invest solely in incremental improvements to existing Earth-orbital capabilities, one must define the alternative *value network* created by deep space operations. To explain, the classical S-curve theory of technological development drives one to anticipate and invest in the next technology to rise from below and eventually surpass the current technology's S-curve(s). As Christensen points out, the classical S-curve theory applies only to *sustaining innovation* under an existing *value network* (set of qualities or capabilities valued by customers).[67] Note that in applying a business analogy to national security, the "customers" are the interests of the United States and its citizens. The S-curve approach is extremely useful within established measures of value (e.g. wartime performance); where one can stay ahead of competitors' incremental improvements. However, it is vulnerable to *disruptive innovation*, where an outside agent finds, develops, and seizes first-mover advantage within an alternative value network that competes with or supplants the existing one. In a strategic military-economic context, the alternative value network in deep space is similar to that of Earth orbit (access, agility, sustainability), but with greater range, autonomy, and energy management. Energy management consists of storage, transmission, forward positioning, and the ability to harvest and use in-situ resources. These values can be further decomposed into functional requirements (e.g., functions the USAF needs to perform in deep space).

Step 3: based on the functional requirements ("values") above, create Design Reference Missions (DRMs) and corresponding capability-based requirements for propulsion and on-board systems. DRMs should be informed by the capability analysis in Section 1, Space Horizons wargame experience, and lessons learned from participation in forums such as ULA's Cislunar Marketplace Workshop and the Tennessee Valley Interstellar Workshop. DRMs should create scalable mission profiles using mature technical parameters as threshold performance measures, with a range of goals and objectives that anticipate both incremental improvements (*sustaining innovation)* and breakthrough propulsion capabilities (*discovery-driven planning*).

Step 4: finally, by applying the existing S-curves to both current and the new alternative value structures, a framework is created for evaluating and prioritizing technology development. While this approach will not be the only method of developmental planning and decision-making, it will hedge against disruptive innovations from adversaries and competitors. The United States Air Force cannot abandon sustaining innovation, but it can shift and balance resources, organizations, and development approaches to maintain strategic advantage in deep space.

Breakthrough Propulsion: Applying Vision and Rigor in Pioneering Research

For those theories and concepts that are potentially revolutionary, yet unproven, Marc Millis offers a thorough approach based on seven years of lessons learned from his leadership of NASA's Breakthrough Propulsion Physics (BPP) Project.[68] Pioneering research is different from technology improvement work because it involves "balancing the vision required to extend beyond established knowledge, along with the rigor required to make genuine progress."[69] Recall that Christensen warns against the trappings of narrowly focused technology improvement work in The Innovator's Dilemma (he calls it *sustaining innovation*) because it is vulnerable to disruptive innovation. In addition to the requisite vision and rigor, the challenges surrounding pioneering research boil down to the need for consistent evaluation in order to strategically allocate limited resources. Again, a multidisciplinary assessment team will be required.

Millis outlines a process using solicitation cycles for short-duration research, with selection criteria specified. In addition to measures of *technical relevance* and *required resources*, possibly the most important selection criteria is *reliability* or *credibility*.[70] To Millis, this means not trying to "judge technical feasibility during proposal review, because that would constitute a research task unto itself. Instead, focus attention on judging if the proposed work will reach a reliable conclusion upon which other researchers and managers can make sound decisions for the future."[†] To help with that judgment, Millis offers four credibility criteria:[71]

[†] Millis et. al., Reliability as a "success criterion even means that a failed concept (test, device, etc.) is still a success if the information gleaned from that failure provides a reliable foundation for future decisions," 681-2.

- Foundations—the source material from which the proposal was based
- Contrast—how the concepts compare to accrued knowledge (self-criticality)
- Testability—if the research is advancing toward a discriminating test
- Results—how likely results will be reliable for future decisions (pro or con)

Another criterion is some measurement of maturity or progress, for which he employs Applied Science Readiness Levels, the highest being equivalent to Technology Readiness Level 1, which is the lowest measure of *technological* progress (e.g. basic principles observed and reported):[72]

Table 2: Applied Science Readiness Levels

Stage 1	General Physics
Stage 2	Critical Issues
Stage 3	Desired Effects
Within each stage, there are 5 levels, each corresponding to a step of the scientific method:	
Step 0	Pre-Science
Step 1	Problem Formulated
Step 2	Data Collected
Step 3	Hypothesis Proposed
Step 4	Hypothesis Tested and Results Reported

Finally, Millis concludes with three broad strategies for success:[73]

1) Breaking down the long-range goals into near-term immediate "go/no-go" research objectives that can each be assessed within one to three years.
2) Devising a numerical means to impartially compare research options and inherently reject non-rigorous submissions.
3) Addressing a diversified portfolio of research approaches.

Applying the methodology pioneered under Millis' leadership at BPP could be as simple as adopting NASA's Innovative Advanced Concepts (NIAC) Program or using it as an entry path to applied technology demonstration programs, such as DARPA, or what the Space Enterprise Consortium (SpEC) is doing for U.S. Air Force Space Command.[74] The SpEC is part of the Secretary of the Air Force's Bending the Cost Curve (BTCC) initiative, addressed in the next section of this study under recommended acquisition and organizational approaches.[75] Regardless of the chosen course of action, there are other breakthrough propulsion concepts and technologies worth examining. Such a program will require a combination of visionary leadership and rigorous management for which the USAF is well suited. As Eric Davis reports to AFRL in his Advanced Propulsion Study, "there is a subset of excellent BPP concepts that are very credible and rigorous, but their modeling or experimental validation work lacks sufficient funding to proceed, or their experimental research presently gives inconclusive results."[76] In this category of breakthrough propulsion, Mach Effect Thrusters and EmDrive represent two leading candidates for immediate and sustained investigation.

Here it is prudent to address investment strategy, or at least the relative cost implications of breakthrough propulsion research. Whereas the minimum cost of taking nuclear thermal propulsion to a zero-power-critical engine demonstration is $150-200 million under three years, definitive tests of METs and EmDrive are an order of magnitude lower at $100-200,000 over 9-12 months.[77] This is not an apples-to-apples comparison in capability (because the breakthrough effects have yet to be proven), but both represent milestones on a continuum of progress. In other words, both present near-term objectives for capabilities on different time horizons—NTP pay-off sooner (more expensive), MET, and EmDrive pay-off later (less expensive). It may prove that METs and EmDrive, as theorized, are either false or offer too little thrust over power to compete with existing solutions, but at least definitive proof would exist, allowing resource allocation elsewhere. Furthermore, their low cost equates to a low-risk approach to capability demonstration by maintaining credibility, avoiding opportunity loss, and possibly creating competitive advantage. To those ends, a diversified portfolio is required, whereby a consistent double-digit percentage of RDT&E funding is fenced from sustaining innovation and directed toward low-cost, high-risk, potentially high pay-off solutions in a "place many bets" scenario. If the bets are well vetted, like the BPP model, then even a null or sub-optimal result is a valuable pay-off in terms of research progress. The BPP project operated for seven years on a total budget of $1.6 million, or approximately $230,000 per year. Whether or not those breakthrough candidates pay-off directly, ground truth constitutes progress, sustaining innovation continues (and may even benefit from spin-off applications), and deep space capability requirements endure.

Assessment Approach Summary

There are several technologies and concepts suitable for the pursuit of deep space propulsion capabilities. Mach Effect Thrusters and EmDrive represent potential breakthroughs in advanced propulsion that offer a strategic opportunity for the USAF at low cost and low risk of investigation. By developing a better understanding of the nature of space-time, gravitation, inertial frames, quantum vacuum, and other fundamental physical phenomena, discovery-driven planning *complements* sustaining innovation, such that the USAF will stay on the forefront of advanced propulsion systems and applications that will revolutionize how the USAF operates in space.

Section 1.3: Technology Strategy

The advancement of propulsion capabilities is not a standalone undertaking; it requires a host of supporting or linked technologies in a chain-link system, each with their own research and development needs. This in itself is a strategic opportunity. The business strategist Richard Rumelt uses Swedish furniture giant IKEA to illustrate the concept of a chain-link system:

- Each of its core activities must be performed with outstanding efficiency and effectiveness (in IKEA terms: design; manufacturing; distribution; inventory; marketing; and sales).
- Core activities must be sufficiently chain-linked that a rival cannot grab business away by adopting only one of them and performing it well.
- Chain-linked activities should form an unusual grouping such that expertise in one does not easily carry over to expertise at the others (a well-designed mix of resources and competencies).

In this way, it is possible "to create constellations of activities that are chain-linked, […] each benefiting from the quality of the other and the whole being resistant to easy imitation."[78] A chain-link system is not synonymous with a system-of-systems approach, but they have in common a core principle that the whole is more than the sum of its components. While these are not new concepts, they should resonate in military terms of creating synergistic effects, a third offset, or operational agility.[79]

In a larger sense, the U.S. Air Force and U.S. military in general are adept at creating chain-link systems. One could say that the entire chain of DOTMLPF elements (Doctrine, Organization, Training, Materiel, Leadership & Education, Personnel, and Facilities) is central to U.S. advantage at all levels of war (strategic, operational, and tactical). However, in terms of developing national power in space—chain-link systems are a strategic opportunity waiting for action, because no single agency or service has the authority or resources to look beyond parochial—or at most, joint—interests in space. Furthermore, commercial industry partners rarely have incentives to create interoperable systems with anything outside of their market share

unless directed by the government. Even then, configurations change so frequently that interface

management and system integration costs soar. Thus one arrives at a host of questions regarding

deep space infrastructure to guide *discovery-driven planning*:

- Will there be an "operating system" in deep space?
- What language(s) will be used for space traffic control?
- What norms, values, and standards of behavior will dominate?
- What are the optimal rendezvous interfaces and interoperability standards?
- What are the optimal communication and navigation protocols?
- What are the optimal locations for depots and transit hubs?
- What common propellant(s) will depots supply?

To answer these questions, homegrown conferences and initiatives have appeared, though

lacking central organization or leadership commensurate with a public-private partnership on the

scale of transcontinental railroads or commercial air travel. ULA initiated a Cislunar

Marketplace Workshop in early 2017, from which the recommendation arose to reach consensus

on these questions before the new administration issues a revised space plan. Organizers of the

Tennessee Valley Interstellar Workshop recently solicited proposals for *seeding infrastructure*:

> Many interstellar mission concepts rely on substantial infrastructure in our solar system to build,
> power, and launch their vehicles. What is seldom addressed, however, is how to begin to build
> that infrastructure, incrementally and affordably. Abstracts are invited that address that gap, with
> an emphasis on defining the first infrastructure missions that (a) can be launched with existing
> spacecraft, (b) provide an immediate utility in space, and (c) are part of a larger plan to extend
> that capability. This encompasses power production and distribution, mining, construction
> material processing, in-space construction, and propellant harvesting and delivery.[80]

While deep space infrastructure planning is going the open-source route of internet protocol

development by the Internet Engineering Task Force, this too is a strategic opportunity to

influence the foundation of chain-link systems and develop national power in space. Another

advantage of a chain-link systems approach is its applicability at different levels of system

planning. Advanced propulsion concepts like METs or EmDrive are part of their own propulsion system of systems that include power management and distribution (PMAD) and attitude determination and reaction control (ADACS). These spacecraft systems will function within a greater suite of in-space capabilities including ISR, mobility, C2, and strike. Furthermore, because beamed-power and/or in-situ resource utilization architectures enable exponentially greater in-space movement and maneuver capability, the United States must engage at the national level as these architectures take shape.

Finally, a chain-link system of systems approach is resilient, because it is not easily hacked, replicated, or "leapfrogged," as competitors are likely to try. China calls this "recombinative innovation," whereby they acquire or replicate foreign technologies and combine them in novel ways to their advantage.[81] Although China is likely to continue the leapfrog approach to gaining parity and advantage, evidence provided by China space expert Stacey Solomone indicates that "China stands on the cusp of taking a lead role in the global space community and transcending the game of technological catch-up."[82] Furthermore, their capabilities are fast approaching their ambitions, of which their space station, on-orbit logistics, and lunar exploration programs are telling.[83] This is due in large part to China's appreciation for the market potential in space and their parallel investment in aerospace R&D centers (such as the one that produced EmDrive results).[84] Another reason is their whole of government approach to space policy and plans, of which U.S. leadership would be wise to take notice.[85] As this section draws toward closure, another strategic concept emerges as particularly relevant: that U.S. pursuit of in-space capabilities not only creates a chain-link advantage, but also offers *proximate objectives* by which to achieve a range of individual capability advantages such as wireless energy transmission, off-Earth energy production, and fast in-space propulsion.

Section 2 – Acquisition and Organizational Approach

"The Unites States is dependent on space for power projection, yet our current space architecture grows increasingly vulnerable."

- *"Fast Space," Air University, 2016*

Deep space propulsion, much like launch vehicle propulsion technologies (both in development and in use) presents a unique challenge to DOD acquisition professionals. This section of the study will address these challenges and present a two-part solution. This first part will involve a brief analysis of current efforts by the DOD and USAF to streamline acquisitions timelines followed by a proposed acquisition model to develop and deploy deep space propulsion technologies while collaborating with agencies and organizations external to the USAF. The second portion discusses a theoretical organization formed and chartered to develop, test, and acquire deep space propulsion technology and include what the organization would potentially look like. Prior to the discussion on the two-part solution however, it must be clearly stated and addressed that without an affordable, reusable, and reliable launch vehicle or family of vehicles, deep space propulsion, along with a myriad of other operations and deployments (both manned and unmanned) in LEO and beyond, will not be viable.

Low cost and reliable access to space represents the precursor to future space operations, regardless of the function, location, or purpose. Currently, the USAF primarily relies on two entities for launch services for National Security Space (NSS) missions. The first company, United Launch Alliance (ULA), a joint venture between Lockheed-Martin and Boeing, offers the Delta IV and Atlas V launch vehicles under the Evolved Expendable Launch Vehicle (EELV) program. These vehicles have afforded the USAF and other agencies and companies, reliability and assured access to space for over ten years and have successfully launched over 118 missions with all payloads delivered to orbit with 100 percent mission success.[86] While the reliability of the Delta IV and the Atlas V vehicles has taken center stage, so have the costs to the USAF for the provided launch services. Recent measures enacted by ULA to reduce costs include the reduction of peripheral suppliers, the reduction of cycle times, the execution of lean manufacturing practices, and the phasing out of the Delta IV launch vehicle (approximately 30% more than the Atlas V). Additional developmental efforts currently underway for ULA include the advance of the Next Generation Launch System (NGLS), referred to as the Vulcan rocket. Regarding the Vulcan, ULA states, "The NGLS offers our customers unprecedented flexibility in a single system. From LEO to Pluto, the single-core NGLS does it all. The simple design is more cost efficient for all customers, whether defense and national security, NASA science and human spaceflight, or commercial."[87] In short, the current and future cost-reduction efforts of the USAF's primary launch provider revolve around in-house, external manufacturing and supply optimization.

Founded in 2002, employing over four thousand employees, and chartered with the task to put humans on Mars, Space Exploration Technologies, or SpaceX, recently made spaceflight history. At approximately 1830L, on 30 March 2017 from Space Launch Complex 39A at Cape Canaveral Air Force Station, FL, Space X successfully launched and landed the first ever previously flown first stage booster of the Falcon 9 rocket. "After successfully launching a satellite toward geosynchronous orbit, the rocket returned to Earth and landed on a remotely piloted platform, known as a drone ship, in the Atlantic Ocean. It was the company's sixth successful landing on a seaborne platform."[88] Founder and Chief Executive Officer (CEO) Elon Musk stated regarding the event, "It shows you can fly and re-fly an orbit-class booster, which is the most expensive part of the rocket; this is ultimately a huge revolution in spaceflight.[89] Chief

Operating Officer (COO) Gwynn Shotwell stated, "Given the goals of SpaceX are to provide space transportation to other planets, we want to make sure whoever we take can come back."[90] The ability to launch, recycle, and reuse first stage boosters represents a revolutionary approach to reducing the costs and cycle times associated with manned and unmanned spaceflight.

Why is this important to the USAF, NASA, the DOD, or other agencies and/or companies? In order to pursue, explore, exploit, and operationalize the space domain, one must be able to reach it in an affordable and reliable manner. The need for this capability has not gone unnoticed. In a 2016 paper chartered by the USAF's Air University (AU) titled "Fast Space: Leveraging Ultra Low-Cost Space Access for 21st Century Challenges," the challenges and near term opportunities are eloquently recognized and discussed. The authors of this piece lay out four key recommendations to facilitate the accelerated development and deployment of ultra-low-cost access to space (ULCATS).

1. **Partner with US commercial firms pursuing ULCATS using DOD's Other Transaction Authority (OTA):** The USAF should assemble a team to pursue the *authority to proceed* with a competition for jointly-funded (cost-shared) prototype OTAs. The full and competition will seek multiple US commercial partners to develop and demonstrate their proposed space systems in collaboration with USAF financial assistance and broader USG technical resources.

2. **Create a purpose-built organization to manage partnerships with commercial ULCATS efforts:** To succeed, the USAF needs to create a purpose-built organization, notionally called the "New Space Development Office" (NSDO), which utilizes innovative acquisition processes and methods. This organization requires a "Fail-Fast, Fail-Forward" culture as to operationally focused cultures where "failure is not an option."

3. **Shape the interagency environment to ease regulatory burdens and lower barriers to entry:** As the principal DOD Space Advisor (PDSA), the SECAF has the broad view of how the alignment of civil, commercial, and national security can benefit comprehensive national power. We recommend the SECAF as PDSA take an active stance in maturing the policy and regulatory environment outside the DOD that can maximize the benefit of high launch rate, rapid-turnaround RLVs and associated on-orbit capabilities.

4. **Integrate consideration of high launch rate rapid turn-around approaches into the Joint requirements and acquisition process:** The current process of requirements and

acquisition does not incentivize building groundbreaking capabilities. We recommend that relevant DOD organizations create initial capability documents (ICDs) that capture the full suite of opportunities provided associated on-orbit capabilities and champion these to the Joint Requirements Oversight Council (JROC).[91]

Could the principles outlined in the above stated requirements and recommendations apply to the acquisition and development of deep space propulsion technologies? Yes, and they must. The following section, part one of the proposed solution, will provide an overview of the general principles required to effectively develop, mature, and procure deep space propulsion technologies.

Section 2.1 – Overarching Acquisition Guidelines and Recommendations

The USAF is no stranger to pursuing developmental technologies for eventual use in future missions and/or vehicles. Technological breakthroughs such as stealth capabilities are examples of such efforts. The first stealth capable aircraft, the F-117, was initially employed in 1989 over Panama, and later in Iraq and Kosovo. The technology enabled the USAF to deliver conventional weapons, effectively executing denial operations without the need for fighter escorts. The aircraft was developed in concert with Lockheed-Martin's Skunk Works, which secured the funding to develop two prototypes with financial assistance from the federal government. Without this partnership, the deployment of this critical and strategic technology could have never happened. Deep space propulsion is no different.

Similar to the development of aircraft technologies, partnerships between DOD, specifically the USAF, governmental agencies, and commercial companies are essential for deep space propulsion technology development and maturation. Recent acquisition reforms have been developed and put into practice in response to budget restrictions recently enacted and levied upon the DOD. Once such effort is the Bending the Cost Curve (BTCC) approach. Dr. Camron

44

Gorguinpor, the director of the Air Force Transformational Innovation Office, Air Force Office of Acquisitions stated, "BTCC is coming up with ideas with industry, then going out and trying those ideas to see if we can actually drive down cost, increase capability, and get it delivered faster."[92] He continues, "Everything we do with BTCC is in collaboration with industry. They are a big part of the solution, so working closely with them helps us come up with better ideas of things that we should be doing."[93] USAF entities such as the Air Force Research Laboratory (AFRL) have enacted processes under the BTCC approach to work with companies who can demonstrate a capability and potentially be on contract to develop and mature the technology in as little as three weeks.[94] "Also included in the BTCC is the Cost Capability Analysis program that would create better transparency by providing more awareness of Air Force requirements to industry to reduce the costs and development times for Air Force systems."[95] BTCC represents a critical opportunity and potential acquisitions vehicle for deep space propulsion technology development as it strives to improve interactions with industry partners and expand competition among traditional and non-traditional industry partners.

Whether under the BTCC approach or not, the key to developing this technology is facilitation of mutually beneficial government and industry partnerships. The previously mentioned AU ULCATS paper shares the same viewpoint, "The USG's traditional acquisition methods are unlikely to achieve ULCATS. Non-traditional partnerships using OTAs have a much higher chance of success. The USAF has the existing authorities it needs for non-traditional partnerships to jump start the virtuous cycle with commercial firms."[96] These partnerships have been extremely successful in the past. Technologies and prototypes such as the Atlas V/Delta IV rockets, the Advanced Short Take-off Vertical Landing, and the Global

Hawk have evolved and been deployed due to strategic partnerships from government and industry.[97]

To establish these partnerships and leverage the synergies gained from them, the USAF must first announce to industry and agencies the need for such technology. This announcement would be similar to a traditional request for proposal (RFP). As concrete requirements for this technology have not been established, this RFP will be general in nature to generate interest among industry partners. Following the RFP, the USAF would sponsor an event where industry can come and present and/or demonstrate their capabilities. The USAF is currently sponsoring similar events. "A PlugFest is a specialized industry event where companies collaborate and demonstrate their existing capabilities in live demonstrations for government customers. However, there is no contracting aspect to a PlugFest. Under our new PlugFest Plus approach, we will put in place a mechanism whereby a vendor could walk away with a contract just a few weeks after an event; we accomplish this by combining these industry events with an Army acquisition model, which minimizes barriers for companies to participate."[98] While the potential exists for a contract award under the current PlugFest construct, such action would be premature in the case of deep space propulsion.

Following the industry event, a down select would be necessary to determine which technologies should be sponsored for further development under other transaction authority (OTA). Unlike traditional acquisitions programs, deep space propulsion will not benefit from a traditional Federal Acquisition Regulation (FAR) based process. Under the OTA arrangement, the government supplies capital, deep technical expertise, and fixed infrastructure beyond the ability of any company to sustain, and the possibility of future purchases if they succeed. The exotic nature of the technology requires a nontraditional approach under the leadership and guidance of individuals willing to collaborate with industry and divorce themselves from

traditional procurement practices with the mission of acquiring deep space propulsion

technology enabling the advancement of the US space presence. In order to divorce themselves

from traditional FAR based acquisitions processes, leadership in this process must support and

defend individuals under them and provide necessary top-cover. The partnership construct

leverages the expertise of government and industry to effectively drive down risk using in-house

expertise rather than an outside source with limited information on the developmental and

acquisition efforts underway. The following section, part two of the proposed solution, will

outline what that organization could look like.

Section 2.2 – The Advanced Propulsion Technology Directorate (SMC/APT)

The solution to the deep space propulsion procurement effort is multifaceted and lies in not

only developing technologies, but also creating an organization that is capable of coordinating

and marshalling the numerous efforts alongside industry partners identified from the previously

mentioned down select. At present, the majority of military payloads are developed through and

acquired at Los Angeles Air Force Base, California, Space and Missile Systems Center

(LAAFB/SMC). Geographically placing the new Advanced Propulsion Technology Directorate

(SMC/APT) at SMC situates the organization among other directorates facing similar challenges

(potential synergies) as well placing the organization across the street from recognized

propulsion experts, the Aerospace Corporation. In addition to placing the directorate amongst

peers, placing the directorate at SMC geographically puts SMC/APT near industry partners such

as Boeing, Lockheed Martin, and the NASA Jet Propulsion Laboratory (JPL), among others.

Additional potential locations for placement could potentially include Edwards AFB, California

(AFRL rocket propulsion directorate) or NASA Marshall Space Flight Center (Huntsville, AL).

Regardless of the location, the newly formed directorate will follow a similar construct and

organization as other directorates at SMC.

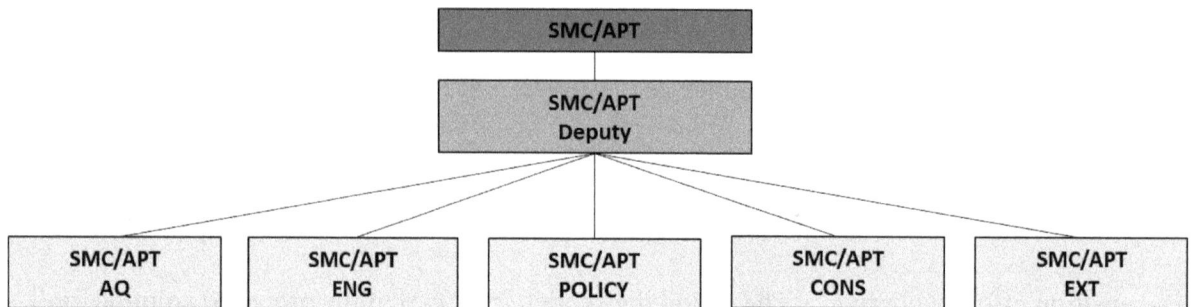

Figure 3: Advanced Propulsion Technology Directorate

A Colonel (O-6) or government civilian (GS) equivalent ideally, would lead SMC/APT. The

deputy billet would require a Lieutenant Colonel (O-5) or government civilian equivalent. The

branches underneath the proposed leadership construct represent the major efforts the

organization will focus on in order to meet the intent and vision; to execute deep space

propulsion technology development, maturation, and deployment under an OTA construct.

The first branch, the acquisition branch, will utilize USAF acquisition professionals, both

active duty and GS, to ensure the integrity of all SMC/APT acquisition efforts. A Major (O-4) or

O-5 with a Defense Acquisition University (DAU) acquisition certification of level three or

higher will lead this branch. The DAU level three certification ensures not only time in the field,

but a knowledge of the process gained from the diverse DAU acquisitions curriculum. As with

other directorates at SMC, the acquisitions branch must be in lockstep with the propulsion

developers and contractors to ensure accountability and must have direct communication with

SMC/APT leadership. A monthly and quarterly contractor feedback mechanism must also be constructed to adjudicate performance and milestone accomplishment.

Similar to the acquisitions branch, the engineering branch will be led by an O-4 or O-5 with a DAU acquisition or engineering certification of level three or higher for much of the same reasons. This individual will also hold the title of Chief Engineer and relay any technical issues to leadership as appropriate. This organization must work closely with the acquisitions branch and play an integral part in the feedback process outlined in the previous paragraph as such technical requirements represent key milestones and deliverables. The engineering branch will largely be comprised of USAF active duty and GS civilians in the first lieutenant (O-2) to captain (O-4) (or GS equivalent range) grades. In addition to these key personnel, propulsion experts from the Aerospace Corporation and NASA would augment the team as permanent staff. The majority of the directorates in SMC utilize these professionals, as they possess a deeper knowledge in the areas of expertise to which they are assigned. The organization facilitates this knowledge by harboring an environment that encourages it and minimizes lateral personnel moves.

The third branch, the policy branch, shall be led by an O-4 or O-5 with a background in political military affairs. The policy branch staff must contain personnel who are intimately familiar with space policy, treaties, and matters involving the legal ramifications of deploying exotic, potentially nuclear, propulsion technologies. Finally, this branch must establish clear lines of communication with the Air Force Space Command (AFSPC) Strategic Plans, Programs, Requirements, and Analysis directorate (AFSPC/A5/8/9) through coordination with SMC senior leadership.

The contracting branch shall be led by an O-5 or GS equivalent with experience in the OTA process and contracting ramifications associated with such action. This branch shall lead the charge on ensuring the contracting effort is clear, concise, efficient, and legal. They shall be integrated into the decision matrices, contractor feedback, policy development, and operational requirements. Incorporating these professionals into the organization from the outset is imperative as it sets the stage for future cooperative contracting efforts.

The fourth and final branch of SMC/APT shall be the liaison and external affairs branch. This branch, due to limited manning levels at SMC and personnel shortages across AFSPC, shall be an additional task executed in conjunction with the primary acquisition, engineering, and policy duties. This approach would serve two purposes; one, it would alleviate the requirement for additional personnel dedicated solely to this function, and two, it would guarantee personnel selected for this mission would have experience in the areas of expertise organic to SMC/APT. This mission of this organization would be to serve as ambassadors, work alongside, and gain insight into the deep space propulsion activities and technology development by external organizations. These organizations include universities, agencies, startups, and hybrids such as the NASA Jet Propulsion Laboratory (JPL). Establishing these relationships would create two-way communication and synergistic relationships between the lead AFSPC agency tasked with propulsion development and additional organizations developing innovative technologies to aid in the effort. Establishing monthly updates and quarterly program reviews (virtual or in person) with representatives from these organizations will prove paramount to the preservation of information crossflow amongst vested parties, potential providers, and SMC/APT professionals.

Section 3 – Strategic Rationale for Deep Space Operations

President Dwight Eisenhower once said, "the mind of man is aroused by the thought of exploring the mysteries of outer space, *and* through such exploration man hopes to broaden his horizons, add to his knowledge, and improve his way of living."[99] Deep space exploration is one of the most important endeavors that civilization will undertake. The forthcoming analysis will provide four reasons why humankind must extend its reach into the solar system, and thus master deep space travel. First, history demonstrates that exploring new frontiers is necessary to generate new waves of technological advancements. The American western expansion and the birth of the space age provide useful examples to illustrate this point. Second, deep space exploration helps to address environmental considerations such as Earth overpopulation, resource constraints, and resource opportunities. Third, deep space capabilities help protect and advance national security interests by providing assured access to space. Assured access to space is particularly important since space is now a contested domain. Lastly, deep space exploration provides unique economic opportunities – these opportunities include space based minerals / resources, and the potential for space based industries. The analysis begins by defining some key terms.

For the purpose of this analysis, it is important to understand the distinction between deep space and outer space. The international community does not agree on a standard definition for the term outer space, but most space law experts agree that outer space can refer to the lowest altitude at which objects can orbit above the Earth, approximately 60 miles.[100] This analysis will primarily refer to deep space operations. Deep space is the portion of outer space that is beyond geosynchronous orbit. In other words, deep space refers to an area of outer space that is further than 22,236 miles above Earth.[101] Using these key definitions, the next portion of the analysis

51

will use historical examples to show how exploring new frontiers generates new waves of technological advancements.

The American western expansion is the first historical example that illustrates the importance of exploring new frontiers—to begin, one must understand what the western expansion was. The story of the United States has always been one of westward expansion, or what Theodore Roosevelt described as "the great leap westward."[102] The journey started on the east coast and continued until it reached the Pacific Ocean. From 1800 to 1900, the United States tripled in size and the geographic distribution of the population shifted from about seven percent living in the west to roughly sixty percent.[103] Manifest destiny, a term coined by John O'Sullivan in 1845, was the guiding principle for the westward expansion. Manifest destiny was the moral obligation for Americans to spread their institutions and liberate people from tyranny.[104] Using this historical context, the analysis will now begin to frame the importance of deep space exploration by identifying the applicable lessons for today.

Similar to the manner in which the American west was unchartered territory, the deep space frontier is largely untapped today—the importance of transportation and innovation are the applicable historical lessons. Transportation is important as a key enabler, but it is also the primary impediment to reaching new domains. The decrease in transportation costs aided the American expansion west.[105] More specifically, the completion of the transcontinental railroad in 1869 is what facilitated the westward expansion. In fact before the transcontinental railroad, the journey across mountains, plains, rivers and deserts was just too risky – most migrants instead chose to travel by sea, taking the six-month route around Cape Horn at the tip of South America.[106] The example shows that while transportation was the main hurdle, it ultimately became the key enabler for reaching the new territories. Today deep space is unchartered territory – transportation is the key enabler, just as it was for the American expansion west. The second lesson is the importance of innovation. The American western expansion ultimately "led to the greatest explosion in innovation the world has ever seen." [107] For example, one-hundred years after this frontier closed, the nation discovered antibiotics; developed nuclear power; brought electricity to cities; invented television and computers; produced telephones; and grew aviation from the Wright Brothers to the safe commercial airline travel.[108] The lesson one should learn from that is, as humankind begins to explore the deep space frontier, the resulting opportunities in innovation will be limitless. Exploring a new frontier forces civilization to reach new heights and to develop new technologies that will enrich the human experience. The next section of the analysis will provide another historical example to show the importance of exploring new frontiers.

The birth of the space age also illustrates the importance of exploring new frontiers – to begin, it is first important to understand what the space age was. For the purpose of this analysis, the space age encompasses the time related to the space race. The space age started on October 4, 1957 when the Soviet Union launched Sputnik.[109] The Soviets had the distinction of putting the first man-made object into space. The Soviet's technological achievement surprised the United States, so the government doubled its efforts to catch up with the Soviets – this marked the beginning of the space race. The United States feared that the Soviets could use their new technology for more "sinister purposes."[110] Eventually *Explorer* became the first US satellite to launch on January 31, 1958, almost four months later. The Soviets went on to achieve a series of other firsts in the 1950s and 1960s: "first man in space, first woman, first three men, first space-walk, first spacecraft to impact the moon, first to orbit the moon, first to impact Venus, and first craft to soft-land on the moon." The United States then took a leap ahead in the 1960s with the

Apollo lunar-landing program. Using this historical context, the analysis will now begin to frame the importance of deep space exploration by explaining the key parallels and important lessons for today.

Just as there was no way to reach space in the early 1950s, there is no efficient (i.e. fast, resourceful, and sustainable) way to reach the deep space frontier today—the importance of transportation and innovation are the important historical lessons from this example as well. Transportation was the key enabler for the beginning of the space age. For the early launch of *Sputnik* and *Explorer,* the Soviets and the Americans used chemically fueled launch vehicles (the *Sputnik PS* and the *Juno 1,* respectively).[111] That solved the problem of reaching space, but the parallel to today is that now there is no efficient means of reaching deep space. Transportation is the key element for deep space travel, just as it was at the beginning of the space age. Another major lesson is the importance of innovation. The ability to reach space ultimately had a profound impact because it drove innovation and technology. The following are some of the noteworthy inventions and capabilities that came out of the space age: intercontinental ballistic missile technology; advancements in robotics; environmental / atmospheric insights; worldwide communication; smoke detectors; cordless tools; enriched baby food; protective paint; scratch resistant glasses; sneaker insoles, etc.[112] NASA argues that if a technology is "cordless, fireproof, automated, or lightweight and strong, there is a good chance it was born during the space age.[113] An unexplored frontier is the deepest driver of innovation that exists, which means exploring the deep space frontier would lead to countless new technologies, just as the beginnings of the space age did. An examination of the space age proves that the question of 'should we explore?' must be seen in deep historical context, not in the context of present-day politics or whims.[114] The next portion of the analysis will begin to discuss how deep space exploration helps to address environmental considerations—it begins by examining Earth overpopulation.

As the Earth becomes more populated, humankind will begin to have geographical limitations (land and area), making deep space operations even more vital. To illustrate this point consider the rate at which the population is growing. As of March 2017, the current Earth population is almost 7.5 billion people and during the 20th century alone, the population in the world has grown by over four billion people.[115] Furthermore, in 1970 there were roughly half as many people in the world as there are now.[116] The foregone conclusion is that Earth will eventually have too many people (some might argue that problem exists today), and that will lead to territory conflicts. Too many people and a finite amount of space is a certain recipe for conflict. The international community will need places to put the growing numbers of people. The alternative could be an overproduction of smog, unbreathable air, a world of skyscrapers, and vegetation / habitats destroyed. Deep space endeavors provide the unique opportunity to become a multi-planet civilization. Elon Musk, founder of SpaceX, has aspirations of building a city on Mars—he argues that if humans become a multi-planet species, humanity is likely to propagate into the future much further than if we are a single-planet species.[117] The population statistics support Musk's belief because without another planet, the future on Earth may become rife with conflict. The next portion of the analysis will continue to discuss how environmental considerations drive the need for deep space exploration – the section examines potential natural resource implications.

The concern over natural resources closely ties to the issue of overpopulation because a finite amount of natural resources with a growing population, further leads to conflict – deep space operations offer a unique opportunity to address these concerns. To illustrate this point, consider that the Earth population needs fertile land (i.e. food), fresh water, energy, and biodiversity to survive. However, the Earth is already experiencing an erosion of farmland; there is overuse of both surface and groundwater; dwindling supplies of finite fossil fuels; escalating extinction of plant and animal species; more than two billion humans are malnourished and experience unhealthy living conditions; and about forty-thousand children die each day from disease and malnutrition.[118] Moreover, around sixty percent of all major cities are at risk of at least one type of major natural disaster,[119] which exacerbates the concern because existing amounts of natural resources could dwindle over time. Space offers a unique opportunity to cultivate new resources, make new discoveries, and it has virtually unlimited resources. Ouyang Ziyuan, chief scientist of China's moon-exploration program, believes that the moon could serve as a "new and tremendous supplier of energy and resources for human beings."[120] Helium 3, which is a potential fuel for fusion power, provides a useful case-in-point. Researchers believe the moon contains one million tons of recoverable Helium 3, enough to power all of Earth for thousands of years.[121] Scientists consider Mars a "resource bonanza,"[122] because although it offers so many unanswered questions, it could be habitable for humans. Asteroids are also extremely useful. The resource content from near-Earth asteroids could sustain 100 billion people, and the materials found in the Asteroid Belt could support 10 quadrillion people.[123] The National Space Society points out that non-terrestrial sources of rare materials may be of great importance here on Earth and the parts of the solar system that are most accessible from Earth (i.e. the Moon, asteroids, and Mars and its moons) are rich in materials that are of great potential value to humanity in deep space.[124] While the resource limitations on Earth present a significant challenge, they also underscore the importance and unique opportunity of deep space exploration. The next portion of the analysis will discuss how deep space operations help protect and advance national security interests by providing assured access to space.

National Security in Deep Space

Space is now a contested domain, which means that the ability to maneuver quickly and efficiently through deep space can also help maintain assured access to space—to begin, it is important to understand how space is a contested. Assured access to space means having sufficiently "robust, responsive, and resilient space transportation capabilities that are available to enable and advance civil and national security missions."[125] John Hyten, former commander of Air Force Space Command, says that with today's national reliance on space capabilities, assured access has gone from important to imperative.[126] Joint Publication 3-14 says that in today's global environment, all segments of a space system are vulnerable to interference or attack—space segments are vulnerable to attacks from direct-ascent antisatellite (ASAT) interceptors, laser blinding, and dazzling; ground-to-satellite link segments are susceptible to jamming and other forms of interference, and ground segments (e.g. launch and command and control) are vulnerable to attack.[127] To demonstrate one such vulnerability, consider that in January of 2007 China deliberately destroyed one of its defunct satellites using a ground based, medium range ballistic missile, proving their ASAT weapon capability.[128] The Joint Operating Environment 2035 points out that future adversaries will have the ability to impede the free operation of satellites and they will use ASAT weapons for kinetic strikes against space assets.[129] The Chinese conducted similar tests in 2010, 2013, and 2014[130] and according to General Jay Raymond, they proved that "soon every satellite in orbit will be able to be held at risk."[131] The Secretary of the Air Force, Deborah James, further characterized the risk when she said, "military commanders have fully realized how fundamental space-based effects have become to every military operation in the world—the problem is that U.S. adversaries recognize it as well."[132] China and Russia continue to move toward military-focused initiatives where they develop weapons explicitly designed to affect America's eyes and ears in space.[133] US forces must be able to address these growing threats. Understanding that space is a contested domain, the next section will explain how deep space capabilities help provide assured access to space.

Developing more efficient deep space travel capabilities supports assured access to space by making it harder for an adversary to target space assets and contest the space domain. To begin, it is important to note that the United States can only access the space environment from a limited number of deniable Earth launch points. Deep space travel capabilities would alleviate these single point limitations and allow the nation to traverse between the Earth and Space domain. There would also be fewer concerns over fuel consumption, and the systems would be faster and likely, more maneuverable.

Space assets currently travel according to a predictable orbital pattern. Part of the reason for that is because the US government designs satellites for a certain "life expectancy,"[134] which means space operators must conserve fuel. Deep space operations would require a more sustainable, renewable, and efficient fuel source. Therefore, by developing deep space travel capabilities, satellite operators would also have the ability to move through space with less concern over fuel consumption. If a ground or space based threat exists, space operators can attempt to evade the threat without concern over losing viable spacecraft life. A space asset that does not have to worry about a finite amount of fuel becomes harder to target because it has more maneuverability options and greater reach throughout the space domain.

In addition to fuel efficiency, deep space assets would also require higher rates of speed. To understand the importance of travel speed, consider that it has taken anywhere from 128-333 days to reach Mars using the existing means of propulsion.[135] A mastery of deep space travel

requires significantly shorter travel timelines, which means that deep space assets would, by

definition, be faster and likely more maneuverable. Once again, speed and maneuverability

make space assets harder to target. Using an example from the air domain, it is easier to attack a

C-5 aircraft than an F-35, because the F-35 is faster, more agile, and highly maneuverable. Deep

space operations require faster, more efficient modes of travel and that capability makes space

assets harder to target. Moreover, it supports assured access to space because these capabilities

ensure a "more robust transport capability."[136] The final portion of the analysis will show how

deep space exploration provides unique economic opportunities—it begins by looking at the

potential for space based minerals and resources.

Deep space travel offers unique economic opportunities because, as scientific evidence

suggests, space is rich in minerals and other resources. The most basic laws of economics are

the law of supply and the law of demand, which says that "the quantity of a good supplied rises

as the market price rises, and falls as the price falls; *and* the quantity of a good demanded falls as

the price rises, and vice versa."[137] Earth has a finite amount of resources—space offers the

opportunity to replenish those resources and find new ones, thereby impacting the universal laws

of supply and demand. Supplying new resources and/or replenishing finite ones will ultimately

affect price and thus, overall economic growth. Planetary scientists argue that the economic

opportunities are limitless. [138]

Consider the following examples: Mars has silica and iron oxide; and asteroids are a good

source of carbon, nitrogen, ferrous metals, phosphates, silicates, and water for cement (if

engineers can mine them).[139] Researchers have also found evidence of water on Mars and the

planet appears to have characteristics and a history similar to Earth's.[140] Of course water is "an excellent source of fuel (especially as liquid hydrogen and liquid oxygen, or when combined with carbon dioxide to form methalox)… it is also important for human habitation, for drinking-water and oxygen to breathe, and to use as radiation shielding or for growing crops."[141]

The moon is another example. The moon is a known source of aluminum[142] and many believe it can serve as a supplier of energy and other resources.[143] The recoverable Helium 3 alone would be worth over $100 trillion dollars.[144] If these nearby celestial bodies offer such resources, the opportunities offered in deep space would seem limitless. Replenishing nonrenewable resources is extremely important for the long-term prosperity of life on Earth, and through deep space operations, researchers would likely identify new types of minerals and resources. Such discoveries would have a tremendous impact on the global community, and they will provide a significant economic advantage. The next section will discuss the potential for space-based industries, and examine how that also provides unique economic opportunities.

A spaced based industry offers another unique economic opportunity because it can expand the means by which a nation produces wealth. There are several examples of how a nation can build its economy through space-based industries. The first example would be space tourism. The Federal Aviation Administration says that space tourism will grow into a one billion dollar industry over the next few years and the National Space Society estimates the industry's size could eventually swell to as high a one trillion dollars.[145] A mastery of deep space travel and technology would allow space tourism to prosper and generate significant wealth for the nation's economy.

Space mining provides a second example. Astrophysicist Neil deGrasse Tyson says, "the first trillionaire there will ever be is the person who exploits the natural resources on asteroids."[146] As previously mentioned, researchers already know that asteroids provide useful minerals, but engineers are unable to effectively mine them. Some space mining companies estimate that the accessible resource content of just a single asteroid could be anywhere from 200 billion to 100 trillion dollars.[147] Only deep space travel can enable such mining. The capabilities developed to support space mining will contribute to vital technologies and provide vital resources.[148]

Another potential space based economy follows the vision of Amazon founder, Jeff Bezos. Bezos has a "vision of 'millions of people living and working in space' and moving heavy industry and energy to space in order to save Earth."[149] Once again, the economic implications of people working in space and moving Earth based industries to space are significant. Not only does that preserve resources, it also supports economic growth by expanding industrial opportunities to a new domain. The opportunities for a space based economy or industry are limitless. The aforementioned examples are only a small subset. Tourism, mining, and expanding Earth based industries into space should be near-term priorities, but other examples might include space trade, producing multiple space stations, and space manufacturing. Deep space operations provide an opportunity to explore new and revolutionary space based industries, which would have a profound impact on the economy.

Former President Barack Obama said the United States' goal is "no longer just a destination to reach… *rather* it is the capacity for people to work and learn and operate and live safely beyond the Earth for extended periods of time, ultimately in ways that are more sustainable and even indefinite."[150] Although the space age began as a race for security and prestige, today the stakes are arguably even higher. Humankind knows so little about the deep space domain, which is why it truly is the final frontier. The nation must be able to efficiently travel further into the solar system and beyond. History supports the notion that societies and technology will advance through deep space operations. Deep space exploration also helps to address some noteworthy environmental concerns and it will serve national security interests for years to come. Lastly, the economic impact of deep space exploration would be profound. All of these areas demonstrate the importance of deep space operations. While it may seem costly and challenging, the deep space domain will chart a new course for history and the United States must lead the way.

Conclusion

"In the 19th century, Admiral Alfred T. Mahan articulated the interactions between naval power and maritime commerce, and sea power's special significance during times of peace. The strategic linkages between space commerce and space power are similar. It was British commercial maritime leadership and innovation that enabled Britain to build the most powerful naval fleet in the world. In the 21st Century, space economic power will extend America's ability to project power during times of peace."
- *"Fast Space," Air University, 2016*

This study began with a definition of the deep space capabilities required for the coming century in space. Like the findings of the ULCATS report, the benefits of fast in-space capabilities go well beyond the military instrument of national power. Collaborating with academia and private industry to advance space commerce while ensuring the security and viability of space lines of communication "will create a strategic situation in which the United States is likely to gain and hold the upper hand."[151] The benefits are commensurate with national leadership, economic power, and political advantage. While international norms and standards of behavior may prevent costly conflict and congestion in space, they will not restrain competition—for knowledge, position, and resources. Likewise, commercial competition in space engenders an advanced space industrial base, bolstering the military's "ability to win and prevent wars by ensuring its freedom of action and superiority of position."[152] By assessing current technologies and breakthrough propulsion candidates against capability-based measures of performance, new opportunities become imperative for the USAF and the nation.

Finally, one recent theory of airpower posited that technology is a unifying element of airpower for three reasons: 1) technology is required to enter the warfighting domains of air, space, and cyber, 2) technology is required to maneuver with or within the domains, and 3) technology is required to generate effects within or from the domains.[153] Furthermore, airpower theory and practice—from the days of Billy Mitchell to the present—show that technology changed the character and conduct of warfare, though perhaps not war's nature. Therefore, this study proceeded from the acknowledgment that technology is a fundamental requirement for creating or improving access, maneuver, or effects within or from the space domain. To maintain position and advantage in space, the USAF should invest in further development of new and emerging in-space propulsion technologies while leading a whole-of-government effort to establish requirements and policy guidance to support deep space operations.

Acknowledgments

Primary thanks go to our advisor, Lt Col Peter Garretson for inspiring, mentoring, and creating the opportunity to expand the USAF's horizons in space, or at the very least, expand our own. The Space Horizons Research Group would not exist without the sponsorship and support of the Air University Commander, Lt Gen Steven Kwast. General Kwast provided impromptu guidance at the outset and encouragement throughout, suggesting we extend the Fast Space concept beyond GEO operations. This suggestion followed his discussion with Brig Gen S. Pete Worden (retired), for whose input we are also grateful. We were introduced to subject matter experts across the Department of Defense, NASA, and commercial space community regarding advanced in-space propulsion technology. We owe great thanks to all of you: Brian Beal, Jean Luc Cambier, Kendra Cook, Mike Elsperman, Jonathan Geerts, Mike Holmes, Justin Koo, Rob Lobbia, John Mankins, Rob Martin, Mike McDonald, Jody Merritt, Greg Meholic, Marc Millis, Jim Peterka, Jess Sponable, Dave Stephens, Travis Taylor, Lance Williams, Marcus Young, and Brent Ziarnick. Finally, we are ever grateful for the support of our classmates and families while students at the Air Command and Staff College. All mistakes herein are our own.

Endnotes

(Notes may appear in shortened form except online sources. For full details, see appropriate bibliographic entry.)

[1] Brig Gen (ret) S. Pete Worden, Chairman, Breakthrough Prize Foundation, to the Commander, Air University, letter, subject: Ultra-Low Cost Access To Space (ULCATS) Study: A Review from Beijing, 11 October 2016.

[2] U.S. Air Force, AFFOC, 11.

[3] Ziarnick, 16-25.

[4] NASA, "Human Exploration of Mars."

[5] Rumelt, 111.

[6] Christensen, 143.

[7] Ibid., 157.

[8] Meyer et al., TA 2-4.

[9] DeRoy and Reed, 6-7.

[10] Beal, 6.

[11] Ibid.

[12] Houts et al., 5.

[13] John Venable, "If Trump Wants Lower F-35 Costs, He Should Compete F135 Engine," *Breaking Defense*, 17 January 2017, http://breakingdefense.com/2017/01/if-trump-wants-lower-f-35-costs-he-should-compete-f135-engine/.

[14] Houts et al., 5.

[15] Lockheed-Martin, "Compact Fusion Research and Development," accessed 6 April 2017, http://www.lockheedmartin.com/us/products/compact-fusion.html.

[16] "Rocket Performance Graph," Project Rho website, accessed 5 April 2017, http://www.projectrho.com/public_html/rocket/enginelist.php.

[17] Bussard, 6.

[18] Ibid., 1.

[19] Lockheed Martin

[20] Meyer et al., TA 2-69.

[21] "Research and Development," Ad Astra Rocket Company, Webster, TX (2017), http://www.adastrarocket.com/aarc/research-and-development.

[22] AcqNotes, "Technology Development: Technology Readiness Level (TRL)," http://www.acqnotes.com/acqnote/tasks/technology-readiness-level.

[23] Meyer et al., TA 2-29.

[24] Scott et al., TA 3-30.

[25] Ibid. For a comprehensive orbital debris mitigation plan, see Major Joshua Wehrle's first Space Horizons study.

[26] Coopersmith, 1.

[27] Lubin et al., 1.

[28] Ibid., 2.

[29] Ibid., 1.

[30] Sercel, 29.

[31] Joel Sercel, PhD, ICS Associates Inc., to the author, e-mail, subject: fast in-space paper, 21 April 2017.

[32] Joel Sercel, "Optical Mining of Asteroids, Moons, and Planets to Enable Sustainable Human Exploration and Space Industrialization," 6 April 2017, https://www.nasa.gov/directorates/spacetech/niac/2017_Phase_I_Phase_II/Sustainable_Human_Exploration.

[33] Sercel, 8.

[34] Ibid., 134.

[35] Woodward, 184.

[36] Beal, 8.

[37] Ibid. Note: the total thrust generated by a Hall thruster is a primarily a function of electromagnetic forces, and to a much lesser degree gasdynamic forces. AFRL researcher Daniel L. Brown synthesized historical models with modern experimentation and analysis to deliver a comprehensive measure for the performance of HETs: "The separation of scalar thrust into mass-weighted and momentum-weighted terms enabled factorization of HET anode efficiency into the product of (1) energy efficiency, (2) propellant efficiency, and (3) beam efficiency." Using beam efficiency as a measure of performance is a highly recommended standard practice going forward.

[38] Sellers, 577.

[39] Gina Anderson and Lori Rachul, "NASA Works to Improve Solar Electric Propulsion for Deep Space Exploration," NASA Release 16-044, 19 April 2016, https://www.nasa.gov/press-release/nasa-works-to-improve-solar-electric-propulsion-for-deep-space-exploration.

[40] Klaus et al.

[41] Brophy et al.

[42] Woodward, 5.

[43] Woodward, 4.

[44] Williams, 92.

[45] Williams, 56 and 267.

[46] Meholic, i.

[47] Williams, 267.

[48] Meholic, i.

[49] Williams, 114-115.

[50] Williams, 102, 111, and 115.

[51] Woodward, ix.

[52] Mary-Ann Russon, "EmDrive: Chinese space agency to put controversial tech onto satellites 'as soon as possible'," International Business Times, 19 December 2016, http://www.ibtimes.co.uk/emdrive-chinese-space-agency-put-controversial-tech-onto-satellites-soon-possible-1596328.

[53] Greene, 332.

[54] Michael McDonald, Ph.D., Naval Center for Space Technology, Naval Research Laboratory, Washington, DC, to the author, e-mail, subject: AFRL - EM Drive Contact, 3 March 2017.

[55] White et al., 11.

[56] Gilster et al., "Uncertain Propulsion Breakthroughs?," 30 December 2016, http://www.centauri-dreams.org/?p=36830.

[57] Taylor, 8.

[58] Marc Millis, Tau Zero Foundation, to Lt Col Peter Garretson, e-mail, subject: Millis at AFIT, 1 November 2016.

[59] Thorne, 29-30.

[60] Davis, 62 and Millis et al., Chapter 18: "On Extracting Energy from the Quantum Vacuum," 569-603.

[61] Millis et al., 423.

[62] Williams, 35.

[63] Gilster.

[64] Taylor, 8.

[65] Meyer et al., TA 2-4.

[66] Marc Millis, Tau Zero Foundation, to the author, e-mail, subject: Millis at AFIT, 4 April 2017. Also see Tennessee Valley Interstellar Workshop (2017), https://tviw.us/submissions/.

[67] Christensen, 39-42.

[68] Millis et al., Chapter 22: "Prioritizing Pioneering Research," 663.

[69] Ibid.

[70] Millis et al., 690.

[71] Ibid., 700.

[72] Ibid., 683.

[73] Ibid., 700.

[74] National Aeronautics and Space Administration, "NIAC Overview," Space Technology Mission Directorate, 16 February 2017, https://www.nasa.gov/content/niac-overview.

[75] U.S. Department of Defense, "Space Enterprise Consortium Other Transaction Request for Information," Federal Business Opportunities, Solicitation Number: FA8814-16-9-0001, 17 October 2016, https://govtribe.com/project/space-enterprise-consortium-other-transaction-request-for-information-1.

[76] Davis, 58.

[77] Millis, e-mail, 1 Nov 2016.

[78] Rumelt, 123.

[79] U.S. Air Force, 7.

[80] Millis, e-mail, 4 Apr 2017. Also see Tennessee Valley Interstellar Workshop (2017), https://tviw.us/submissions/.

[81] Solomone, 37.

[82] Ibid., 31.

[83] "China progresses toward first Orbital Logistics Mission, Space Station Module finishes Assembly," 4 March 2017, and "Chang'e 3 Mission Overview," SpaceFlight101.com, http://spaceflight101.com/change/change-3/.

[84] Juan et al.

[85] Solomone, 26.

[86] "Quick Facts," accessed on 13 March 2017, http://www.ulalaunch.com.

[87] Ibid.

[88] Wattles, Jackie, "Space X makes history: it launched a used rocket and then landed it in the ocean," *CNN*, 31 March 2017, http://money.cnn.com/2017/03/30/technology/spacex-launch-ses-10-reused-rocket/.

[89] Ibid.

[90] Ibid.

[91] Fast Space, 4.

[92] Secretary of the Air Force Public Affairs, "Securing the Future by Bending the Cost Curve," 3 November 2015, http://www.af.mil/News/Article-Display/Article/627140/securing-the-future-by-bending-the-cost-curve/.

[93] Ibid.

[94] Ibid.

[95] Ibid.

[96] Fast Space, 28.

[97] Ibid.

[98] Jim Garamone, "James: New acquisition initiatives aims to cut costs," 15 January 2015, http://www.af.mil/News/Article-Display/Article/560219/james-new-acquisition-initiative-aims-to-cut-costs/.

[99] "The National Space Policy of the United States" on The Office of Space Commerce website, http://www.space.commerce.gov/policy/national-space-policy/. Each President since Eisenhower has defined a national space policy. Eisenhower was a trailblazer of sorts and since making these statements, each president has tried to ensure that as we continue to pursue of new frontiers and address any present-day space challenges.

[100] Rupert Anderson, *The Cosmic Compendium-Space Law*, (Lulu Press, 2015), 1. The author surmises that although the international community has not reached agreement, the term's stated meaning is widely understood.

[101] Elizabeth Howell, "What is a Geosynchronous Orbit," *Space.com*, 24 April 2015, http://www.space.com/29222-geosynchronous-orbit.html. To expand on this definition, a geosynchronous orbit is an orbit that allows satellites to match Earth's rotation. It is located at 22,236 miles (35,786 kilometers) above Earth's equator, a position that is a valuable spot for monitoring weather and enabling communications.

[102] "Westward Expansion Facts" on HistoryNet website, http://www.historynet.com/westward-expansion. The story of the Americas is incomplete without a full understanding of "the Great Leap Westward." Roosevelt coined this term to describe the monumental impact of fully exploring this new frontier.

[103] University of Southern California, "What Caused Westward Expansion In The United States?," *ScienceDaily*, 29 February 2008, www.sciencedaily.com/releases/2008/02/080228150402.htm.

[104] "Westward Expansion Facts" on HistoryNet website, http://www.historynet.com/westward-expansion

[105] University of Southern California, "What Caused Westward Expansion In The United States?" *ScienceDaily*, 29 February 2008, www.sciencedaily.com/releases/2008/02/080228150402.htm. The importance of transportation is a profound parallel to today's deep space conundrum. The parallel suggests that with access to and within the deep space domain, advancement never takes place. If the Americans could not efficiently reach the west, the resulting levels of innovation may have been profoundly different.

[106] "Transcontinental Railroad" on History.com, http://www.history.com/topics/inventions/transcontinental-railroad. The journey west was long and tedious. The author continues to say that some migrants would risk yellow fever and other diseases by crossing the Isthmus of Panama and traveling via ship to San Francisco. At a minimum, the trip would take six months to make the long journey across the country.

[107] "The Importance of Exploration" on National Aeronautics and Space Administration (NASA) website, https://www.nasa.gov/missions/solarsystem/Why_We_01pt1.html. NASA points out how valuable it was for the early explorers to sail west and the fact that those early exploratory missions changed the course of history.

[108] Ibid. When mankind ventures into an unexplored place, it begins to drive innovation. New technologies develop, many times out of necessity, to adapt to new settings and surroundings.

[109] "The Space Age Begins" on NASA website, https://www.nasa.gov/multimedia/imagegallery/image_feature_927.html. The Soviets had the distinct honor of being the first sovereign nation to launch an artificial satellite into orbit around Earth. There is wide agreement in the international community that the Sputnik launch spurred the space "race" or "age." Sputnik weighed 183 pounds and orbited the Earth in roughly 98 minutes.

[110] Paul Dickson, *Sputnik: The Shock of the Century* (Walker and Company Press: New York, 2001), 3-5. "1957 Sputnik Launched" on History.com, http://www.history.com/this-day-in-history/sputnik-launched. "Sinister purposes" is a great depiction of what truly caused the panic in America. Paul Dickson explains how the United

States was in utter shock over the launch of Sputnik. The *history.com* account summarizes the sentiment of the time—what will the Soviets do with such technology?

[111] "Explorer-I and Jupiter-C: The First United States Satellite and Space Launch Vehicle" on NASA website, https://history.nasa.gov/sputnik/expinfo.html. Juno used liquid oxygen, as oxidizer, and "Hydyne" (60% unsymmetrical, dimethylhydrazine and 40% diethylenetriamine), as fuel. Sputnik used a kerosene and LOx mixture. Both launch vehicles used chemical propellant, which is still predominant today.

[112] Ker Than, "The Top 10 Revelations of the Space Age," *Space.com*, 26 October 2010, http://www.space.com/23-top-10-revelations-space-age.html.

[113] "The Importance of Exploration" on NASA website, https://www.nasa.gov/missions/solarsystem/Why_We_01pt1.html.

[114] Ibid. NASA contends with the perception that scientific missions are a "nice-to-have." NASA leadership argues that we must think of deep space exploration as imperative–too important to be relegated to simple political interest.

[115] "Worldometers" on Worldometers website, http://www.worldometers.info/world-population/. Counters have been licensed for the United Nations, BBC News, and World Expo. Data provided through United Nations Population Division, World Health Organization (WHO), Food and Agriculture Organization (FAO), International Monetary Fund (IMF), and World Bank.

[116] Ibid.

[117] Ajai Raj, "Elon Musk: SpaceX wants to Build a City on Mars," *Business Insider,* 12 September 2014, http://www.businessinsider.com/elon-musk-wants-to-build-a-city-on-mars-2014-9.

[118] David Pimentel, X. Huang, A. Cordova and M. Pimentel, *Impact of a Growing Population on Natural Resources: The Challenge for Environmental Management Frontiers,* (The Interdisciplinary Journal of Study Abroad), 1.

[119] Rachel Nuwer, "Is the World Running Out of Space," *BBC News,* 1 September 2015, http://www.bbc.com/future/story/20150901-is-the-world-running-out-of-space.

[120] Peter Garretson, "Guess What Could Be Totally Missing From the New U.S. President's Intel Briefing: Oh, just the entire outer-space thing," War is Boring, 29 September 2016, https://warisboring.com/guess-what-could-be-totally-missing-from-the-new-u-s-presidents-intel-briefing-9ce4881643e5#.jvt6lr1rq.

[121] Steve Almasy, "Could the moon provide clean energy for Earth," *CNN,* 21 July 2011, http://www.cnn.com/2011/TECH/innovation/07/21/mining.moon.helium3/.

[122] John S. Lewis, Mildred Shapley Matthews, and Mary L. Guerrieri, Resources of Near-Earth Space, (The Arizona Board of Regents), Part IV. The title "resource bonanza" aptly illustrates why the community at large is so excited about Mars, which offers many unique opportunities and may provide answers to questions about Earth's evolution.

[123] John Lewis, "Cutting the Umbilical Cord to Earth," February 2016. Lewis is Professor Emeritus of Planetary Sci., LPL and Chief Scientist at Deep Space Industries.

[124] Lewis et al., Abstract.

[125] "National Space Transportation Policy," on NASA website, https://www.nasa.gov/sites/default/files/files/national_space_transportation_policy_11212013.pdf. The policy goes on to say that the capacity to provide assured access resides "partly within the United States Government and, increasingly, within the U.S. private sector, which offers space transportation services and capabilities for the United States Government and commercial applications."

[126] Torri Ingalsbe, "Air Force Focuses on assured access to space," on af.mil, 30 April 2015, http://www.af.mil/News/ArticleDisplay/tabid/223/Article/587048/air-force-focuses-on-assured-access-to-space.aspx. General Hyten's words highlight that the United States can ill afford to enter institutional self-interest battles. In others words, this is not the time for "turf-battles." The United States must ensure assured access to space, and when facing an adversary who challenges the space domain, the nation must be postured to respond and it must be able to maintain on-orbit capabilities.

[127] U.S. Joint Chiefs of Staff, Joint Publication (JP) 3-14: Space Operations (Washington, DC: Government Printing Office, May 2013), V-7.

[128] Mike Gruss, "US Official: China Turned to Debris-free ASAT tests Following 2007 Outcry," spacenews.com, 11 January 2016. http://spacenews.com/u-s-official-china-turned-to-debris-free-asat-tests-following-2007-outcry/ The Chinese action was widely condemned throughout the international space community primarily because it left a cloud of potentially hazardous debris in orbit, a significant problem for any user of the space domain.

[129] U.S. Joint Chiefs of Staff, *Joint Operating Environment (JOE) 2035* (Washington, DC: Government Printing Office, 14 July 2016), 32-33.

[130] Andrea Shalal-Esa, *"*Pentagon cites new drive to develop anti-satellite weapons," Reuters World News, 8 May 2015, http://in.reuters.com/article/pentagon-satellites-idINDEE9460FQ20130507. Also see Mike Gruss, "US Official: China Turned to Debris-free ASAT tests Following 2007 Outcry," spacenews.com, 11 January 2016,

http://spacenews.com/u-s-official-china-turned-to-debris-free-asat-tests-following-2007-outcry/. Although it did not create the debris that the first incident did, follow on testing demonstrated their capability to destroy on-orbit assets.

[131] Colin Clark, "Chinese ASAT Test was successful: Lt Gen Raymond," Breakingdefense.com, 14 April 2015, http://breakingdefense.com/2015/04/chinese-asat-test-was-successful-lt-gen-raymond/. General Raymond was the commander of 14th Air Force and his statement highlights the growing realization that our satellites are vulnerable to attack. Given US dependency on satellites and their vulnerability, senior policy leaders must react.

[132] Ingalsbe, "Air Force Focuses on assured access to space." The following is the full excerpt from the article. James stated assured access to space must remain the nation's top priority going forward, especially with increased threats and potential adversaries within the space arena: "While our combatant and theater commanders have fully realized how fundamental space-based effects have become to every military operation in the world, our potential adversaries have been watching and working to challenge those very capabilities."

[133] Steven Jiang, "China: We will be on Mars by the end of 2020," *CNN,* 5 January 2017, http://www.cnn.com/2016/12/28/asia/china-space-program-white-paper/. US adversaries aim to "deny, degrade, deceive, disrupt, or destroy" America's eyes and ears in space.

[134] Owen Kurtin, "Satellite Life Extension: Reaching for the Holy Grail," *SatelliteToday,* 1 March 2013, http://www.satellitetoday.com/publications/2013/03/01/satellite-life-extension-reaching-for-the-holy-grail/. "The useful lifetime of geosynchronous orbit satellites averages about fifteen year—a limit primarily imposed by the exhaustion of propellant aboard. The propellant is needed for "station-keeping," which means maintaining the satellite in its orbital slot and in-orbit orientation, or attitude, so that its antennae and solar panels are properly pointed. When the propellant is nearly exhausted, the satellite reaches the end of its active life and must be moved to a graveyard orbital slot, even though the satellite's other systems and payload are often in working order."

[135] Nola Taylor Redd, "How Long Does It Take to Get to Mars?" *Space.com*, 13 February 2014, http://www.space.com/24701-how-long-does-it-take-to-get-to-mars.html. It took Mariner 7 128 days and Viking 2 roughly 333 days. Fuel conservation is the primary factor.

[136] "National Space Transportation Policy," on NASA website, https://www.nasa.gov/sites/default/files/files/national_space_transportation_policy_11212013.pdf.

[137] Al Ehrbar, "Supply," *The Library of Economics,* http://www.econlib.org/library/Enc/Supply.html.

[138] Jane Grant, *Leaving Earth,* (University of Plymouth Press, Plymouth England, 2016). Ian Crawford is a prominent scientist from the University of London. He points out the resource rich environment that other planets may offer, including new elements that have yet to be discovered, as well as vital elements needed here on Earth.

[139] Wired Staff, "The 12 Greatest Challenges for Space Exploration," *Wired.com,* 16 February 2016, https://www.wired.com/2016/02/space-is-cold-vast-and-deadly-humans-will-explore-it-anyway/. This short list of known information about nearby orbital bodies illustrates how little we know about the rest of the universe.

[140] "Why we Explore–Human Space Exploration" on NASA website, https://www.nasa.gov/exploration/whyweexplore/why_we_explore_main.html. NASA continues to say that "Mars has characteristics and a history similar to Earth's, but we know that there are striking differences that we have yet to begin to understand. Humans can build upon this knowledge and look for signs of life and investigate Mars' geological evolution, resulting in research and methods that could be applied here on Earth." A mission to our nearest planetary neighbor provides the best opportunity to demonstrate that humans can live for extended, even permanent, stays beyond low Earth orbit.

[141] Bruce Dorminey, "Deep Space Industries To Probe Near Earth Asteroids," *Forbes.com*, 18 November 2016, https://www.forbes.com/sites/brucedorminey/2016/11/18/deep-space-industries-to-probe-near-earth-asteroid/#5cb189885e3b. Daniel Faber is the CEO of Deep Space Industries, a Silicon Valley based company. He argues that water is just one of the natural resources that can be mined in space and that has significant implications for Earth and the future of civilization.

[142] Wired Staff, "The 12 Greatest Challenges for Space Exploration," *Wired.com,* 16 February 2016, https://www.wired.com/2016/02/space-is-cold-vast-and-deadly-humans-will-explore-it-anyway/.

[143] Garretson, "Guess What Could Be Totally Missing…" The author Lt Col Peter Garretson, quotes Quyang Ziyaun, the chief scientist of China's moon-exploration program. China is devoting an abundance of resources to moon exploration and deep space operations, because they recognize the potential global impact of this new domain.

[144] Paul Spudis, *The Value of the Moon: How to Explore, Live, and Prosper in Space Using the Moon's Resources,* (Smithsonian Books: Washington DC, 2016).

[145] Katie Kramer, "Build the economy here on Earth by exploring space," *CNBC,* 3 May 2015, http://www.cnbc.com/2015/05/01/build-the-economy-here-on-earth-by-exploring-space-tyson.html.

[146] Ibid., Tyson is arguably the most famous astrophysicist in America and he goes on to say that an engineer who comes out with a new patent to take you to a place, intellectually, physically, that has never been reached before, those become the engines of tomorrow's economy.

[147] Ram Jakhu, Joseph N. Pelton, and Yaw Out Mankata Nyampong, *Space Mining and Its Regulation,* (Springer Praxis Books: Switzerland, 2017), 3, and Kristen Bobst, "What is an asteroid is worth?," *Mother Nature Network*, 15 March 2016, https://www.usatoday.com/story/tech/nation-now/2017/01/18/nasa-planning-mission-asteroid-worth-10000-quadrillion/96709250/. Other sources include Elizabeth Howell, "'Trillion-Dollar Asteroid' Zooms by Earth as Scientists Watch," *Space.com*, 28 July 2015, http://www.space.com/30074-trillion-dollar-asteroid-2011-uw158-earth-flyby.html, and Matthew Reynolds, "Infoporn: where to mine asteroids to get the deep space dollars," *Wired.com*, 31 October 2016, http://www.wired.co.uk/article/infoporn-mining-for-asteroids, and Rob Waugh, *DailyMail.com*, 21 May 2012, http://www.dailymail.co.uk/sciencetech/article-2147404/Found-The-single-asteroid-thats-worth-60-billion-years-financial-output-entire-WORLD.html#ixzz4gBnaDP3A.

[148] Ibid., 12.

[149] Garretson, "Guess What Could Be Totally Missing…"

[150] "The National Space Policy of the United States" on The Office of Space Commerce website, http://www.space.commerce.gov/policy/national-space-policy/. It is fitting to end this discussion on "why we need deep space" with a quote from the most recent President to establish Space Policy. The analysis started with President Eisenhower who was a trailblazer for all Presidents who followed him, and concludes with words from President Obama.

[151] Fast Space, 32.

[152] Ibid., 31.

[153] Deaile, 5.

Bibliography

Brophy, John R., Louis Friedman, Nathan J. Strange, Thomas A. Prince, Damon Landau, Thomas Jones, Russell Schweickart, Chris Lewicki, Martin Elvis, and David Manzella. "Synergies of Robotic Asteroid Redirection Technologies and Human Space Exploration," 2014.

Beal, Brian. The Air Force Research Laboratory's In-Space Propulsion Program. No. AFRL-RQ-ED-TP-2015-008. Air Force Research Laboratory, Edwards AFB, CA, Aerospace Systems Directorate, 2015.

Bussard, Robert W. "The Advent of Clean Nuclear Fusion: Superperformance Space Power and Propulsion." In *57th International Astronautical Congress*, 2006.

Brown, Daniel L., C. William Larson, Brian E. Beal, and Alec D. Gallimore. "Methodology and historical perspective of a Hall thruster efficiency analysis." Journal of Propulsion and Power 25, no. 6 (2009): 1163-1177.

Christensen, Clayton. *The Innovator's Dilemma: When New Technologies Cause Great Firms to Fail*. Boston, MA: Harvard Business Review Press, 2013.

Coopersmith, Jonathan C., and Eric Davis. "A Strategic Roadmap for Commercializing Low-Cost Beamed Energy Propulsion Launch Systems." AIAA SPACE 2016, p. 5555, 2016.

Davis, Eric W. "Advanced Propulsion Study." AFRL-PR-ED-TR-2004-0024. Las Vegas, NV: Warp Drive Metrics, 2004.

Deaile, Mel. "A Theory of Airpower." Lecture, Air Command and Staff College, Maxwell AFB, AL, 28 March 2017.

DeRoy, Rich S. and John G. Reed. "Vulcan, ACES, and Beyond: Providing Launch Services for Tomorrow's Spacecraft." Guidance, Navigation, and Control, Advances in the Astronautical Sciences, AAS-16-052, 2016.

Dewar, James A., and Robert Bussard. *The Nuclear Rocket: Making Our Planet Green, Peaceful and Prosperous*. Burlington, Ontario: Apogee Books, 2009.

"Fast Space: Leveraging Ultra Low-Cost Space Access for 21st Century Challenges," Air University, Maxwell AFB, AL, December 2016.

Funaro, Gregory V., and Reginald A. Alexander. "Technology Alignment and Portfolio Prioritization (TAPP): Advanced Methods in Strategic Analysis, Technology Forecasting and Long Term Planning for Human Exploration and Operations, Advanced Exploration Systems and Advanced Concepts," 2015.

Garretson, Peter A. "Asteroid Strike! Asteroid Mining! Will the Air Force Have a Role?" *Air & Space Power Journal* 31, no. 1 (Spring 2017).

Greene, Brian. *The Fabric of the Cosmos: Space, Time, and the Texture of Reality*. New York: Vintage Books, 2004.

Herman, Daniel A., Walter Santiago, Hani Kamhawi, James E. Polk, John Steven Snyder, Richard R. Hofer, and J. Morgan Parker. "The Ion Propulsion System for the Solar Electric Propulsion Technology Demonstration Mission." National Aeronautics and Space Administration (2015).

Houts, Mike, Sonny Mitchell, Ken Aschenbrenner, and Anthony Kelly. "Development and Utilization of Space Fission Power and Propulsion Systems." National Aeronautics and Space Administration (2017). https://ntrs.nasa.gov/archive/nasa/casi.ntrs.nasa.gov/20170002040.pdf

Juan, Yang, Wang Yuquan, Li Pengfei, Wang Yang, Wang Yunmin, and Ma Yanjie. "Net thrust measurement of propellantless microwave thrusters." *Acta Physica Sinica (in Chinese)*, Chinese Physical Society, 2012.

Klaus, Kurt K., M. S. Elsperman, and F. Rogers. "Mission Concepts Enabled by Solar Electric Propulsion and Advanced Modular Power Systems." In *AAS/Division for Planetary Sciences Meeting Abstracts*, vol. 45. 2013.

Lubin, Philip, Gary Hughes, Johanna Bible, and Isabella Johansson. "Directed Energy For Relativistic Propulsion and Interstellar Communications." Journal of the British Interplanetary Society 68, no. 5/6 (2015): 172.

Mankins, John C. *The Case for Space Solar Power*. Norfolk: Virginia Edition Publishing, 2014.

Meholic, Greg, Lance Williams, and Mark Crofton, "Aerospace Review of Mach Effect Propulsion Research and Current Experiments (Final Report)," AEROSPACE REPORT NO. TOR-2014-03848, 31 December 2014. Distribution Statement F: Further dissemination only as directed by SMC/AD or higher DOD authority.

Meyer, Mike, L. Johnson, F. Chandler, D. Coote, H. Gerrish, D. Goebel, B. Palaszewski, T. Smith, and H. White. "NASA Technology Roadmaps TA 2: In-space Propulsion Technologies." National Aeronautics and Space Administration, Office of the Chief Technologist, 2015.

Millis, Marc G., and Eric W. Davis, eds. *Frontiers of Propulsion Science*. Vol. 227. Reston, VA: American Institute of Aeronautics & Astronautics, 2009.

NASA Architecture Steering Group. "Human Exploration of Mars, Design Reference Mission 5.0." National Aeronautics and Space Administration, Reference number: SP-2009-566, July 2009.

Parissenti, G., N. Koch, D. Pavarin, E. Ahedo, K. Katsonis, F. Scortecci, and M. Pessana, "Non-Conventional Propellants for Electric Propulsion Applications," Space Propulsion 2010-1841086, (2010). http://aero.uc3m.es/ep2/docs/publicaciones/ahed10a.pdf

Rumelt, Richard P. *Good Strategy/Bad Strategy: The Difference and Why it Matters*. New York: Crown Business, 2011.

Scott, John, C. Mercer, F. Chandler, G. Carr, M. Houts, C. Iannello, T. Lawrence, S. Surampudi, and C. Taylor. "NASA Technology Roadmaps TA 3: Space Power and Energy Storage." National Aeronautics and Space Administration, Office of the Chief Technologist, 2015.

Sellers, Jerry Jon. *Understanding Space: An Introduction to Astronautics*. 3rd ed. St Louis, MO: McGraw-Hill, 2005.

Sercel, Joel C. "Stepping Stones: Economic Analysis of Space Transportation Supplied From NEO Resources." Progress Report on Grant No. NNX16AH11G, Winter 2016.

Solomone, Stacey. *China's Strategy in Space*. New York: Springer Books, 2013.

Taylor, Travis S. "A Concept for Producing Very Large Propulsive Forces using High-Q Asymmetric High Energy Laser Resonators." Presented at the Directed Energy Symposium. Huntsville, AL, February 2017.

Thorne, Kip. *The Science of Interstellar*. New York: WW Norton & Company, 2014.

U.S. Air Force, "Air Force Future Operating Concept: A View of the Air Force in 2035," Washington, DC: Government Printing Office, September 2015.

Williams, Lance L. and Heidi Fearn, eds. *Estes Park Advanced Propulsion Workshop: Scheduled Technical Proceedings*. Manitou Springs, CO: Konfluence Press, 2017

Woodward, James F. *Making Starships and Stargates: The Science of Interstellar Transport and Absurdly Benign Wormholes*. New York: Springer Science & Business Media, 2012.

Ziarnick, Brent. *Developing National Power in Space: A Theoretical Model*. Jefferson, NC: McFarland & Company, Inc., 2015.